Cambridge Elements ≡

Elements in the Structure and Dynamics of Complex Networks
edited by
Guido Caldarelli
Ca' Foscari University of Venice

MULTILAYER NETWORK SCIENCE

From Cells to Societies

Oriol Artime
Complex Multilayer Networks Lab, Fondazione Bruno Kessler

Barbara Benigni
Complex Multilayer Networks Lab, Fondazione Bruno Kessler

Giulia Bertagnolli
Complex Multilayer Networks Lab, Fondazione Bruno Kessler

Valeria d'Andrea
Complex Multilayer Networks Lab, Fondazione Bruno Kessler

Riccardo Gallotti
Complex Multilayer Networks Lab, Fondazione Bruno Kessler

Arsham Ghavasieh
Complex Multilayer Networks Lab, Fondazione Bruno Kessler

Sebastian Raimondo
Complex Multilayer Networks Lab, Fondazione Bruno Kessler

Manlio De Domenico
*Complex Multilayer Networks Lab, Fondazione Bruno Kessler and
University of Padova*

CAMBRIDGE
UNIVERSITY PRESS

CAMBRIDGE
UNIVERSITY PRESS

University Printing House, Cambridge CB2 8BS, United Kingdom

One Liberty Plaza, 20th Floor, New York, NY 10006, USA

477 Williamstown Road, Port Melbourne, VIC 3207, Australia

314–321, 3rd Floor, Plot 3, Splendor Forum, Jasola District Centre,
New Delhi – 110025, India

103 Penang Road, #05–06/07, Visioncrest Commercial, Singapore 238467

Cambridge University Press is part of the University of Cambridge.

It furthers the University's mission by disseminating knowledge in the pursuit of
education, learning, and research at the highest international levels of excellence.

www.cambridge.org
Information on this title: www.cambridge.org/9781009087308
DOI: 10.1017/9781009085809

© Oriol Artime, Barbara Benigni, Giulia Bertagnolli, Valeria d'Andrea, Riccardo
Gallotti, Arsham Ghavasieh, Sebastian Raimondo, and Manlio De Domenico 2022

First published 2022

A catalogue record for this publication is available from the British Library.

ISBN 978-1-009-08730-8 Paperback
ISSN 2516-5763 (online)
ISSN 2516-5755 (print)

Multilayer Network Science

From Cells to Societies

Elements in the Structure and Dynamics of Complex Networks

DOI: 10.1017/9781009085809
First published online: August 2022

Oriol Artime
Complex Multilayer Networks Lab, Fondazione Bruno Kessler
Barbara Benigni
Complex Multilayer Networks Lab, Fondazione Bruno Kessler
Giulia Bertagnolli
Complex Multilayer Networks Lab, Fondazione Bruno Kessler
Valeria d'Andrea
Complex Multilayer Networks Lab, Fondazione Bruno Kessler
Riccardo Gallotti
Complex Multilayer Networks Lab, Fondazione Bruno Kessler
Arsham Ghavasieh
Complex Multilayer Networks Lab, Fondazione Bruno Kessler
Sebastian Raimondo
Complex Multilayer Networks Lab, Fondazione Bruno Kessler
Manlio De Domenico
Complex Multilayer Networks Lab, Fondazione Bruno Kessler and University of Padova

Authors for correspondence: Oriol Artime, oartime@fbk.eu;
Manlio De Domenico, manlio.dedomenico@unipd.it

Abstract: Networks are convenient mathematical models to represent the structure of complex systems, from cells to societies. In the past decade, multilayer network science – the branch of the field dealing with units interacting in multiple distinct ways, simultaneously – was demonstrated to be an effective modeling and analytical framework for a wide spectrum of empirical systems, from biopolymer networks (such as interactome and metabolomes) to neuronal networks (such as connectomes), from social networks to urban and transportation networks. In this Element, a decade after the publication of one of the most seminal papers on this topic, the authors review the most salient features of multilayer network science, covering both theoretical aspects and direct applications to real-world coupled/interdependent systems, from the point of view of multilayer structure, dynamics, and function. The authors discuss potential frontiers for this topic and the corresponding challenges in the field for the future.

Keywords: multilayer networks, complex systems, multiplex networks, multilevel networks, interdependent systems

ISBNs: 9781009087308 (PB), 9781009085809 (OC)
ISSNs: 2516-5763 (online), 2516-5755 (print)

Contents

1 Introduction

What is a complex system? It is a *network* of actors or units related by special types of interactions that, together, form a whole. Whether involving two proteins within a cell or two individuals within a social group, relationships and interactions tie these units together in such a way that "the whole is larger than the sum of its parts," a concept initially introduced by the Greek philosopher Aristotle and later exploited by Gestalt psychologists, at the end of the nineteenth century, to explain human perception beyond the traditional atomistic view.

In fact, the "whole" exhibits features that each actor or unit, in isolation, does not and could not. Therefore, it is usually difficult, if not impossible, to understand a system from the analysis of its components alone, as in atomistic or other reductionist theories [309]. The framework required to study such relationships and interactions is known as network science.[1]

The foundations of network science can be found in the pioneering work of Leonhard Euler in 1736, when the famous mathematician provided the first mathematically grounded proof to definitively solve the problem of the Seven Bridges of Königsberg. He mapped the empirical problem of traversing the city of Königsberg – under the constraint that one should use each one of its seven bridges only one time – onto the abstract problem of performing a special walk through a graph. After Euler's solution, graph theory quickly developed in the successive two centuries, culminating in the groundbreaking contributions by Paul Erdős and Alfréd Rényi on random graphs and their statistical analysis at the end of the 1950s.

For decades, social scientists and (systems) biologists have widely used graph theory to map connections between individuals and biological units, respectively, to gain novel insights about the *properties of a system*, the relevance of a *unit within the system*, and the *organization of units within the system*. In 1974, François Jacob, the winner of the 1965 Nobel Prize in Physiology or Medicine, described biology as a science effectively dealing with systems within systems [280], well before the age of genomics and large-scale biology. He recognized that biological systems also can be mapped onto units of systems at a larger scale: in fact, proteins interact with each other to make cells function, cells interact with each other to construct tissues and organs, which in turn interact with each other to build an organism. Finally, at the top of this hierarchical web of interactions, organisms interact with each other to define a population, like our society. In the same decade, similar ideas regarding the

---

[1] We refer the reader to this interesting, nontechnical, and recent introduction to the basic concepts characterizing complex systems [96].

nontrivial interdependencies between scales were laid out by the 1977 Nobel Laureate in Physics, Phillip Anderson, in the context of natural sciences [14].

Social scientists, as biologists, were among the first to face the existence of multiple levels (or scales) as well as multiple *layers* of descriptions for the units of a social system. In the early 1970s, Wayne W. Zachary observed the interactions within a group of individuals belonging to a karate club over three years [304] in order to understand the dynamics of conflicts, which allowed him to predict the outcome of the group split that happened later. He annotated interactions across eight distinct contexts, from "the association in and between academic classes at the university" to "attendance at intercollegiate karate tournaments held at local universities." However, at that time, the mathematical framework required to study a network with multiple *layers of complexity* – such as the eight contexts – was not yet developed and Zachary opted for an approximation: aggregating the multiple interactions across contexts into a single representative number denoting the intensity of the relationship between a pair of actors.

In this work, we seek to better understand the challenges faced by systems biologists and social scientists between the 1970s and the past decade, while introducing the basic concepts required to define the framework of *multilayer network science* with an interdisciplinary language that should be familiar to biologists, social scientists, computer scientists, applied mathematicians, and physicists. Therefore, it will become clear that, for instance, Zachary's approach was a possible model to study the karate club network, but likely neither the most accurate nor the most predictive one. We will discuss under which conditions a system admits a multilayer representation, providing examples such as the ones shown in Figures 1 and 2, where units are individuals and geographic areas, respectively, and interactions represent coauthorship of scientific papers and transportation routes, respectively. Another emblematic example, accounting for the temporal and socio-spatial interdependence typical of many systems, concerns the organization of ecological systems [225]. Finally, very recently, multilayer modeling in systems biology and medicine has been used to integrate information about biological processes, drug targets, genotype, and phenotype to the subset of the human interactome targeted by SARS-CoV-2, the COVID-19 virus [288] (see Figure 3). This work is full of examples like these, and we hope to make clear the broad spectrum of potential interdisciplinary applications of the multilayer framework.

Our ultimate goal is to guide the reader through the potential applications of multilayer modeling, which nowadays provides a well-established paradigm for the analysis of systems characterized by multiple levels and layers of

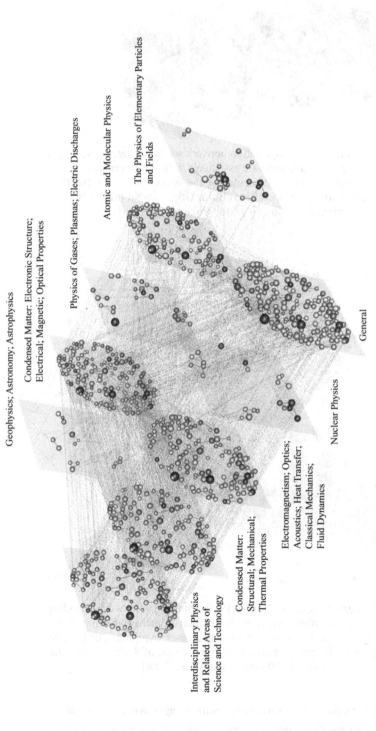

Figure 1 Multilayer representation of a coauthorship network. Nodes represent authors publishing papers in the journals of the American Physical Society, and links connect two authors if they have published a paper together. Layers encode distinct subtopics of physics (e.g., geophysics or nuclear physics). Links within the same layer represent coauthorship of one paper about the same topic, while links between layers indicate coauthorship of one paper categorized simultaneously across distinct topics. Figure from [93] under Creative Commons Attribution-ShareAlike 4.0 International License.

Geophysics; Astronomy; Astrophysics

Condensed Matter: Electronic Structure; Electrical; Magnetic; Optical Properties

Physics of Gases; Plasmas; Electric Discharges

Atomic and Molecular Physics

The Physics of Elementary Particles and Fields

General

Nuclear Physics

Electromagnetism; Optics; Acoustics; Heat Transfer; Classical Mechanics; Fluid Dynamics

Condensed Matter: Structural; Mechanical; Thermal Properties

Interdisciplinary Physics and Related Areas of Science and Technology

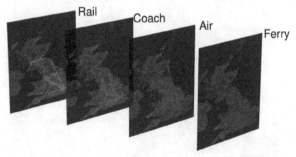

Figure 2 A multilayer transportation network where connections using a particular means of transport are associated with intralayer links and intermodal exchanges are represented by the interlayer links. Here, the national public transportation network for Great Britain [125, 126] as rendered by MuxViz [101]. Figure from [126].

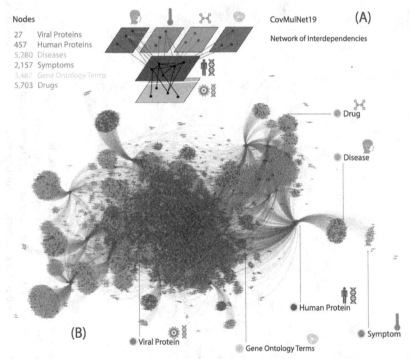

Figure 3 Illustration of CovMulNet19, the multilayer network encoding COVID-19 genotype-phenotype-drug interactions. A schematic map of intra- and interlayer interdependencies between diseases, symptoms, drugs, gene ontology terms, human proteins, and viral proteins of SARS-CoV-2, the COVID-19 virus. Figure from [288].

description, including systems whose structure changes over time. The aim is to provide the reader with the tools required to model and analyze systems in terms of coupled layers, as well as with the conditions under which this approach is plausible.

It is worth remarking here that this work should be considered as an extended introduction to the field but not the most complete one. For this reason, we point the reader to the first reviews [36, 53, 103, 161, 295] and recent books [51, 83] on this topic or more specifically on analysis and visualization of multilayer networks [94], which, taken together with our work, will provide a more comprehensive view of the field.

In Section 2, we will introduce the representation of multilayer networks based on the tensorial formulation [97], providing the mathematical ground for the analytical techniques for structure (Section 3) and dynamics (Section 4), allowing the reader to find a reference for the analysis of versatility (or multilayer centrality) and mesoscale organization (or community detection), as well as for percolation, synchronization, competition, and modeling of intertwined phenomena. Toward the end (Section 5), we will discuss a few selected advances in network science – namely the latent geometry of a complex network based on network-driven processes and the statistical theory of information dynamics leading to the formalism of network density matrices – and their recent generalization and application to multilayer networks. Finally (Section 6), we will show how multilayer networks are ubiquitous and can be used for modeling complex systems, from cells to societies.

2 Representation of Multilayer Systems

2.1 Tensorial Representation of a Complex Network

One convenient way to mathematically represent a complex network is by means of its adjacency matrix [30, 31, 114, 173, 203]. However, to deal with multilayer networks, it might be more convenient to introduce first the more general concept of the tensor, a multilinear function that maps objects defined in a vector space into other objects of the same type, regardless of the choice of a coordinate system. For instance, a simple scalar x is also a rank-0 tensor, a vector x_i is a rank-1 tensor, and a matrix X_{ij} is a rank-2 tensor. More generally, given a vector space $\mathcal{V}$ with algebraic dual space[2] $\mathcal{V}^\star$ over the real numbers $\mathbb{R}$, we can define the tensor M as the multilinear function

$$M: \mathcal{V}^\star \times \mathcal{V}^\star \times \dots \mathcal{V}^\star \times \mathcal{V} \times \mathcal{V} \times \dots \mathcal{V} \longrightarrow \mathbb{R}, \qquad (2.1)$$

---

[2] This is the space of all the possible linear transformations that map an object of $\mathcal{V}$ into a real number. For instance, think about $\mathcal{V} = \mathbb{R}^2$ and the linear functional $f: \mathbb{R}^2 \longrightarrow \mathbb{R}$: it follows that $f(x, y) = ax + by$, with a, b two integer numbers, is an element of $\mathcal{V}^\star$.

where the number of products is m for the vector space and n for its dual. This definition formally characterizes a rank-mn tensor $M^{i_1 i_2 \ldots i_n}_{j_1 j_2 \ldots j_m}$ that is m-covariant and n-contravariant. In fact, under a change of basis B, m components transform as the same linear mapping of the change of basis (B), whereas n components transform as the inverse one (B^{-1}). Therefore, in general, there are two types of canonical basis: the covariant basis denoted by $e_i(a)$ $(a = 1, 2, \ldots, m)$, which is defined in $\mathcal{V}$, and the contravariant (or dual) basis denoted by $e^i(b)$ $(b = 1, 2, \ldots, n)$, which is defined in $\mathcal{V}^\star$. If the vector space is Euclidean, the coordinates of the canonical vectors and their duals are the same, whereas this is not the case in general. In the following, to define an adjacency matrix, or a rank-2 adjacency tensor, we will work in the Euclidean space but we will keep the covariant and contravariant notation, since it will allow us to generalize the results to the case of non-Euclidean spaces. The interested reader can find more about the tensorial framework in any good linear algebra textbook, while for the purpose of this work it is sufficient to understand how we can use tensors in practice in a few key situations.

Let us start by better defining the canonical vectors in the case of networks. For a graph with N nodes, the canonical covariant vectors $e_i(a)$ defined in the space of nodes $\mathbb{R}^N$ are N rank-1 tensors of dimension N with all entries equal to 0 except for the a-th entry, which is equal to 1. Similarly for canonical contravariant vectors. The product of canonical vectors gives canonical matrices – for example, $E_{ij}(ab) = e_i(a)e_j(b)$ is a rank-2 covariant tensor with all components equal to 0 except for the one corresponding to the a-th row and the b-th column, equal to 1. Similarly, we can build contravariant tensors and mixed tensors – that is, tensors obtained by the product between the covariant and contravariant vectors.

The careful reader has noticed at this point that we have defined the *outer* product of two canonical vectors, also known as the Kronecker product, which gives a rank-2 tensor as a result. This result is general: the outer product of two tensors X and Y is a new tensor Z with a number of covariant (contravariant) indices given by the sum of the number of covariant (contravariant) indices of X and Y. Therefore, the outer product of two tensors is always a tensor of higher order than the original ones – for example, $X^k_{ij} Y^{mn}_l = Z^{kmn}_{ijl}$.

It is possible to define also an *inner* product: in this case, we talk about a contraction because the rank of resulting tensor is reduced by two units. For instance, this is the case in the product $X^k_{ij} Y^{mn}_k = Z^{mn}_{ij}$, where the index k is covariant for X and contravariant for Y. This operation corresponds to summing over the components of X and Y identified by the index k. The careful reader has noticed that we have omitted the summation symbol: this choice – known as

Einstein summation convention – is optional and often adopted for simplicity. In the following, we will make use of this convention.

At this point, we are ready to define the adjacency tensor of a complex network in terms of canonical vectors [97] as

$$W_j^i = \sum_{a,b=1}^{N} w_{ab} e^i(a) e_j(b) = \sum_{a,b=1}^{N} w_{ab} E_j^i(ab), \qquad (2.2)$$

where w_{ab} is a real number, usually nonnegative, used to encode the intensity of the interaction between nodes a and b, while $E_j^i(ab) \in \mathbb{R}^{N \times N}$ are the mixed canonical rank-2 tensors. We might wonder if W_j^i is a true tensor, or just a matrix. To this end, it is enough to understand how it transforms under a change of basis

$$B_j^i = \sum_{a=1}^{N} e^{\prime i}(a) e_j(a), \qquad (2.3)$$

a linear function that transforms the basis vector set $\{e^i(a)\}$ into a second set $\{e^{\prime i}(a)\}$. By noting that w_{ab} must be invariant with respect to the change of basis, we have:

$$W_l^{\prime k} = \sum_{a,b=1}^{N} w_{ab} e^{\prime k}(a) e_l^{\prime}(b) = \sum_{a,b=1}^{N} w_{ab} B_i^k e^i(a) e_j(b) (B^{-1})_l^j$$

$$= B_i^k \left[\sum_{a,b=1}^{N} w_{ab} e^i(a) e_j(b) \right] (B^{-1})_l^j = B_i^k W_j^i (B^{-1})_l^j \qquad (2.4)$$

– that is, the adjacency object W_j^i transforms like a tensor [102]. This result is important since a tensor is an object with features that, in general, are not shared by a matrix or, at higher orders, a hypermatrix. In fact, the components of a tensor can always be arranged into hypermatrices, while the opposite is not necessarily true.

Since we work in the Euclidean space, we might wonder why we use this notation and not a simpler one. In general, this is convenient because of the presence of directed relationships between nodes: to distinguish between incoming and outgoing directions, it is sufficient to map this information into covariant and contravariant indices in such a way that the adjacency tensor W_j^i represents a linear transformation that maps nodes into a function of their incoming or outgoing flow. For instance, node a is represented by $e_i(a)$ in the space of nodes and $W_j^i e_i(a) = w_j(a)$ provides a rank-1 tensor encoding the set of nodes linked by a, while $W_j^i u_i = s_j$, with u_i the rank-1 tensor with all components equal to 1, provides a rank-1 tensor encoding the outgoing strength of all nodes. Similarly, $W^i_j e^j(a) = w^i(a)$ gives the set of nodes linking to a, while $W^i_j u^j = s^i$ gives the incoming strength of nodes.

Before moving to the next section, it is useful to define some tensors used throughout this work. We have just seen the rank-1 1–tensor in action: similarly we can define the rank-2 1–tensor $U^i_j = u^i u_j$ or higher-order tensors. Another fundamental tensor is the Kronecker one, defined by δ^i_j, with components equal to 1 if $i = j$ and equal to 0 otherwise.

2.2 Tensorial Representation of a Multilayer Network

In the previous section, we introduced the fundamental procedure required to build an adjacency tensor to represent a classical network (a monoplex). Using a similar procedure, we can build a *multilayer adjacency tensor* to represent a multilayer network, as shown in Figure 4. A multilayer system is characterized by N physical nodes interacting in L distinct ways simultaneously. Each type of interaction defines a *layer*. At variance with single-layer networks, there are more edge sets to encode: as many as the number (L) of layers and, in general, as many as the number $(L(L-1))$ of directed pairwise connections between layers, since we have to specify which node i in a layer α is connected to which node

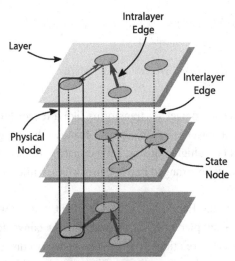

Figure 4 A system where nodes are characterized by three distinct types of interactions, encoded by colored layers. Overall, the system is a multilayer network because to describe relationships we need to specify more than one network. Units are physical nodes: each one is a set of *state nodes* or *replicas*, each one encoding the identity of the corresponding physical node in each layer separately. *Intralayer* edges define connectivity within each layer, whereas *interlayer* edges define connectivity across layers. Reproduced with permission from [93].

j in a layer β $(i,j = 1,2,\ldots,N, \alpha,\beta = 1,2,\ldots,L)$.[3] Note that, for simplicity, we are indicating with Greek letters the indices related to layers and with Latin letters the indices related to nodes.

There are different types of multilayer networks depending on the presence or absence of links between layers and on the way nodes are defined (see Figure 5). In the following, we will mostly deal with the class of systems characterized by interlayer connectivity since it is not possible to define a meaningful multilayer adjacency tensor for the class of edge-colored multigraphs.[4]

Let us introduce the canonical rank-1 vectors $e^\alpha(p)$ $(\alpha,p = 1,\ldots,L)$ in the space of layers $\mathbb{R}^L$, and the corresponding canonical rank-2 tensors $E_\beta^\alpha(pq) = e^\alpha(p)e_\beta(q)$, similarly to what we have done for monoplexes. It is straightforward to show [97] that the linear combination of

$$M_{j\beta}^{i\alpha} = \sum_{a,b=1}^{N} \sum_{p,q=1}^{L} w_{ab}(pq)e^i(a)e_j(b)e^\alpha(p)e_\beta(q) \tag{2.5}$$

fully characterizes a multilinear object in the space $\mathbb{R}^{N \times L \times N \times L}$. This object is, in fact, the desired multilayer adjacency tensor since, under a change of coordinates, it transforms like a tensor:

Multilayer Networks

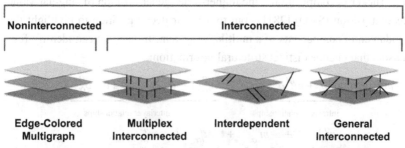

Figure 5 Multilayer networks include a broad spectrum of possible models. Edge-colored networks are useful models when interlayer connectivity is not well defined: this is the case of a social network where edges can represent different types of social relationships (e.g., trust, family, business, etc.) [73, 100, 205]. Conversely, in interconnected networks, interlayer connectivity is well defined and allows us to model a variety of systems [98, 102, 214, 232], including those with interdependencies where nodes control and/or are controlled by nodes in another network [68, 129, 230, 236, 289]. Reproduced with permission from [93].

---

[3] This simple observation suggests that a good candidate for multilayer adjacency tensor should be a rank-4 tensor.

[4] Note that, instead, it is possible to define a valid hypermatrix encoding this object, and this hypermatrix can be thought of as an array of matrices [49].

$$M_{j\beta}^{ri\alpha} = \sum_{a,b=1}^{N} \sum_{p,q=1}^{L} w_{ab}(pq)B_k^i e^k(a)(B^{-1})_j^l e_l(b)\tilde{B}_\gamma^\alpha e^\gamma(p)(\tilde{B}^{-1})_\beta^\delta e_\delta(q)$$

$$= B_k^i \tilde{B}_\gamma^\alpha M_{l\delta}^{k\gamma}(B^{-1})_j^l(\tilde{B}^{-1})_\beta^\delta. \tag{2.6}$$

By indicating with $E_{j\beta}^{i\alpha}(ab;pq) = E_j^i(ab)E_\beta^\alpha(pq)$ the canonical rank-4 tensors, we can simply reduce the definition of the multilayer adjacency tensor to

$$M_{j\beta}^{i\alpha} = \sum_{a,b=1}^{N} \sum_{p,q=1}^{L} w_{ab}(pq)E_{j\beta}^{i\alpha}(ab;pq), \tag{2.7}$$

where $w_{ab}(pq)$ encodes the intensity of the interaction between node a in layer p and node b in layer q. Note that $w_{ab}(pp)$ indicates the weights of the links in layer p.

It is worth noticing that, as for the space of nodes, in the space of layers, we can define multilayer 1–tensors and Kronecker tensors as $U_{j\beta}^{i\alpha} = U_j^i U_\beta^\alpha$ and $\delta_{j\beta}^{i\alpha}$, respectively. Another important tensor, representing a complete multilayer network without self-edges, will be used later in this work to characterize multilayer triadic closure: for consistency, we prefer to introduce it here as $F_{j\beta}^{i\alpha} = U_{j\beta}^{i\alpha} - \delta_{j\beta}^{i\alpha}$.

At this point, the reader should be familiar enough with tensors to note that different decompositions are possible. Here, we are not referring to operations like Tucker decomposition – the higher-order generalization of singular value decomposition (SVD) [281] – but to a linear decomposition to highlight the fundamental components of a multilayer system. In fact, we can identify four tensors that encode distinct structural information:

$$m_{i\alpha}^{j\beta} = \underbrace{m_{i\alpha}^{j\beta}\delta_\alpha^\beta\delta_i^j + m_{i\alpha}^{j\beta}\delta_\alpha^\beta(1-\delta_i^j)}_{\text{intralayer relationships}} + \underbrace{m_{i\alpha}^{j\beta}(1-\delta_\alpha^\beta)\delta_i^j + m_{i\alpha}^{j\beta}(1-\delta_\alpha^\beta)(1-\delta_i^j)}_{\text{interlayer relationships}}$$

$$= \underbrace{m_{i\alpha}^{i\alpha}}_{\text{self-relationships}} + \underbrace{m_{i\alpha}^{j\alpha}}_{\text{endogenous}} + \underbrace{m_{i\alpha}^{i\beta}}_{\text{exogenous}} + \underbrace{m_{i\alpha}^{j\beta}}_{\text{intertwining}}$$

$$= \mathbb{S}_{i\alpha}(M) + \mathbb{N}_{i\alpha}^j(M) + \mathbb{X}_{i\alpha}^\beta(M) + \mathbb{I}_{i\alpha}^\beta(M). \tag{2.8}$$

Here, the components of the tensor are indicated by $m_{i\alpha}^{j\beta}$ ($i,j = 1,2,\ldots,N$ and $\alpha,\beta = 1,2,\ldots,L$), while δ_i^j and δ_α^β indicate the Kronecker delta function in the space of nodes and layers, respectively. The four tensors encode the following relationships:

- **Intralayer interactions**:
    - **self-interactions** ($\mathbb{S}$): from a node to itself;
    - **endogeneous interactions** ($\mathbb{N}$): between distinct nodes belonging to the same layer;

- **Interlayer interactions**:
  - **exogenous interactions** ($\mathbb{X}$): between distinct nodes belonging to distinct layers; and
  - **intertwining** ($\mathbb{I}$): from a node to its replicas in other layers.

Equation (2.8) characterizes the "structural $\mathbb{SNXI}$ decomposition" of the multilayer adjacency tensor M: the models for interconnected systems described in Figure 5 can be characterized by which $\mathbb{SNXI}$ components contribute to the tensor. In particular, we identify:

- **Interconnected multiplex networks**: type $\mathbb{SNI}$;
- **Interdependent networks**: type $\mathbb{SNX}$; and
- **General multilayer networks**: $\mathbb{SNXI}$.

As previously mentioned, edge-colored networks do not admit a meaningful representation in terms of multilayer adjacency tensor, but according to this classification they would define type $\mathbb{SN}$.

From an operational perspective, working with tensors might be complicated and cumbersome. Nevertheless, from a theoretical perspective, the tensorial formulation allows us to write complex equations in a very compact and handy way, and it can be used to guide our intuition about generalizing existing network descriptors for multilayer analysis, as we will see in the next sections. One widely adopted approach is based on the operation known as *matricization* (or *flattening*) [164], which maps a complex object like a high-order tensor into a lower-order object while preserving the information content (see Figure 6 for an emblematic example). In the case of the multilayer adjacency tensor $M_{j\beta}^{i\alpha}$, which is defined in the space $\mathbb{R}^{N\times L\times N\times L}$, this operation corresponds to flatten entries into a rank-2 object defined in the space $\mathbb{R}^{NL\times NL}$, as shown in

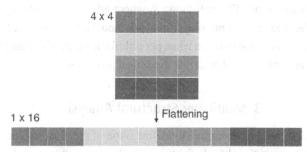

4 x 4

1 x 16   ↓ Flattening

Figure 6 A rank-2 tensor defined in $\mathbb{R}^{4\times4}$ is flattened into a rank-1 tensor defined in $\mathbb{R}^{1\times16}$, without loss of information. Reproduced with permission from [93].

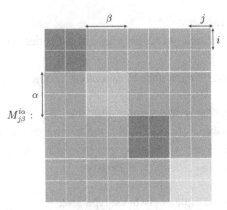

Figure 7 As in Figure 6 but applied to a multilayer adjacency tensor $M_{j\beta}^{i\alpha}$ in $\mathbb{R}^{2\times4\times2\times4}$ flattened into a supra-adjacency matrix in $\mathbb{R}^{8\times8}$. Colors highlight the diagonal blocks where adjacency matrices corresponding to layers are encoded. Off-diagonal blocks encode interlayer connectivity. Reproduced with permission from [93].

Figure 7. It is manifest that there is no loss of information, although care is needed when dealing with this new object, which is known in the literature as a *supra-adjacency matrix* [97, 136, 258].

Working with a supra-adjacency matrix comes with several computational advantages and notational disadvantages. For practical purposes, the flattening allows for a visual inspection of the multilayer adjacency tensor, as shown in Figure 8. The supra-adjacency matrix is in fact a block matrix with intralayer connectivity encoded into diagonal blocks and interlayer connectivity encoded into off-diagonal ones (see also Figure 8). However, it is worth remarking that this is a matter of convention: in fact, this arrangement of blocks is not unique and other arrangements are also valid, although some are less convenient than others for the design of algorithms. This nonuniqueness of the supra-adjacency matrix makes it less suitable for theoretical calculations but still an alternative to higher-order tensors. Figure 9 shows the supra-adjacency matrix representation for three widely used multilayer network models. In the next sections we will use the tensorial formulation when possible for the analysis of the topology of multilayer networks and to define dynamical processes.

3 Multilayer Structural Analysis

In this section, we will operationally define some of the most important theoretical and computational tools for the analysis of a multilayer network. We start with walks (Section 3.1) to define distinct types of connected components (Section 3.2) and we will quickly move to describe several measures of

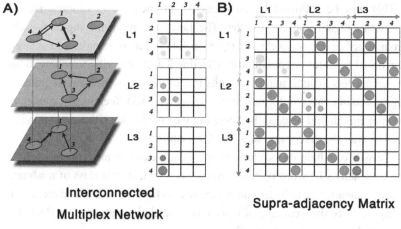

Figure 8 A multilayer network, consisting of three layers, flattened into a supra-adjacency matrix [97, 136]. (A) The interconnected multiplex network (left) with $N = 4$ nodes and $L = 3$ layers; the latter are color coded. Connectivity is weighted (see edge thickness) and directed (see arrows): the adjacency matrix of each layer is shown on the right-hand side of this panel. (B) Result of the flattening procedure: the original rank-4 tensor is now encoded into a rank-2 supra-adjacency matrix with a block structure. Intralayer connectivity is encoded into diagonal blocks (according to colors), whereas interlayer interactions are represented as off-diagonal blocks. Reproduced with permission from [28].

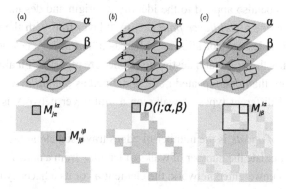

Figure 9 Supra-adjacency matrix representation of three distinct types of multilayer networks [97, 136], where layers are encoded by colors: (a) edge-colored multigraph (no interlayer connectivity), (b) interconnected multiplex network (diagonal coupling in the off-diagonal blocks), and (c) a general interconnected case where interlayer connections are not restricted to replicas (exogenous interactions). Latin letters denote nodes while Greek letters are used for layers. $D(i; \alpha\beta)$ indicates the intensity of the connection between state nodes (i, α) and (i, β). Figure from [46].

node importance within the system – that is, multilayer versatility, which is the generalization of the concept of node centrality (Section 3.3).

This will be followed by an overview of methods to characterize the mesoscale organization of a system (Section 3.4), by identifying how nodes group together into small clusters (Section 3.4.1) such as triangles and in larger functional modules or communities (Section 3.4.2). We will show how to use these concepts to understand how units in multilayer systems are integrated or segregated in terms of information flow (Section 3.4.3). Technically, connected components should be described here but we opted to keep the corresponding section after the description of walks for simplicity.

We will conclude this section by discussing existing methods to quantify the correlations between pairs of layers (Section 3.5): this class of analytical tools is fundamental to better understand results from versatility and mesoscale analysis, since the existence of layer-layer correlation patterns is reflected in structural measures based on walks and paths.

3.1 Basic Definitions

In a generic network, a *walk* is defined as a sequence of adjacent nodes and edges visited by a hypothetical walker. In general, a given edge (but also node) can be traversed more than once, but it is possible to apply some restrictions to a multilayer walk to specify special walks. For instance, we can define a *multilayer trail* as a walk where links can be traversed only one time. A further restriction can be also applied to the identity of origin and destination nodes, and we can define a *multilayer path* as a multilayer trail with the restriction that repeated nodes are not allowed, whereas a *multilayer cycle* is a closed trail where only the origin and destination nodes are repeated. Finally, a closed trail with more than one repeated node is defined as a *multilayer circuit*. An illustration of different types of walks on a multilayer network is shown in Figure 10.

The *length of a walk* is the number of edges traversed along the walks. It is possible to calculate the number of walks of length ℓ from a node i to any other node j. For an unweighted network, the element A_l^k of its adjacency matrix is 1 if there is an edge from k to l and 0 otherwise. Then, if there is a walk with $\ell = 2$ from i to j via k, the product $A_k^i A_j^k$ will be 1. By generalizing this argument, we can write the rank-2 tensor encoding information about walk length between any pair of nodes in the network as the ℓ-th power of the rank-2 adjacency tensor representing the network:

$$\mathcal{W}_j^i(\ell) = (A_j^i)^\ell = A_{j_1}^i A_{j_2}^{j_1} \ldots A_j^{j_{\ell-1}}. \tag{3.1}$$

The same formalism can be used in a weighted network by defining the weight of a walk as the product of the weights of the traversed links. The entries of $\mathcal{W}_j^i(\ell)$ will give the sum of weights of the walks of length ℓ connecting the corresponding pair of nodes.

Walk	Trail	Path	Cycle	Circuit

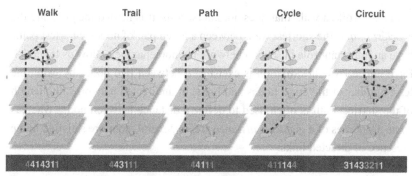

| 414311 | 443111 | 44111 | 411144 | 31433211 |

Figure 10 Illustration of distinct walks possible on a multilayer network. By definition, a walk in a multilayer system is the most general way to move between its layers, nodes, and links. To define walks characterized by distinct features, it is sufficient to restrict the number of times nodes or links can be traversed or to put constraints on origin and destination nodes. The illustration shows a sequence of nodes and edges corresponding to each type of walk (multilayer walk, trail, path, cycle, and circuit) with a dashed line: note that the sequence of nodes visited by each type of walk is explicitly reported below the corresponding multilayer network. Reproduced with permission from [93].

An analogous approach is used to calculate the walk length for multilayer networks. If $M_{j\beta}^{i\alpha}$ is the rank-4 adjacency tensor representing the system, then the entries of the ℓ-th power of this tensor provide the number of multilayer walks of length ℓ between a node i in layer α and a node j in layer β:

$$\mathcal{W}_{j\beta}^{i\alpha}(\ell) = M_{j_1\beta_1}^{i\alpha} M_{j_2\beta_2}^{j_1\beta_1} \ldots M_{j\beta}^{j_{\ell-1}\beta_{\ell-1}}. \tag{3.2}$$

The aggregate representation of multilayer networks allows us to aggregate layers to obtain a network where the number of edges or the weight of an edge is the number of different types of edges between a pair of nodes [82, 97]. It is possible to highlight the topological difference between multilayer networks and their aggregated representations by using the aforementioned formalism. Let $\bar{G}_j^i = M_{j\beta}^{i\alpha} U_\alpha^\beta$ be the aggregate network that accounts for interlayer links: the corresponding rank-2 walk tensor is then given by

$$\begin{aligned}
\bar{\mathcal{W}}_j^i(\ell) = (\bar{G}_j^i)^\ell &= \bar{G}_{j_1}^i \, \bar{G}_{j_2}^{j_1} \ldots \bar{G}_j^{j_{\ell-1}} \\
&= M_{j_1\beta_1}^{i\alpha} U_\alpha^{\beta_1} M_{j_2\beta_2}^{j_1\beta_1} U_{\beta_1}^{\beta_2} \ldots M_{j\beta}^{j_{\ell-1}\beta_{\ell-1}} U_{\beta_{\ell-1}}^\beta \\
&= \underbrace{\left(M_{j_1\beta_1}^{i\alpha} M_{j_2\beta_2}^{j_1\beta_1} \ldots M_{j\beta}^{j_{\ell-1}\beta_{\ell-1}} \right)}_{\mathcal{W}_{j\beta}^{i\alpha}(\ell)} \underbrace{\left(U_\alpha^{\beta_1} U_{\beta_1}^{\beta_2} \ldots U_{\beta_{\ell-1}}^\beta \right)}_{U_\beta^\alpha L^{\ell-1}},
\end{aligned} \tag{3.3}$$

showing that the number of walks with length ℓ between two nodes in the aggregate network is not a linear function of the number of walks of length ℓ between the same pair of nodes in the multilayer network.

In a network, a walk that does not intersect itself is named the *path* and the *shortest path* is the shortest walk between a given pair of nodes. Shortest paths have an important role in several network phenomena: they allow us to model, for instance, how information is exchanged between two nodes by using the least number of traversed nodes and links. In undirected networks, the length of the shortest path is often used to define a distance between nodes whereas some properties of a *geodesic distance* are, in general, no more satisfied in the presence of directed links.

3.2 Connected Components

To identify clusters of nodes that can exchange information in a network, it is useful to analyze the connected components [30, 203].[5] Components are defined as separate parts of the network – that is, a subset of nodes with at least one path between any origin/destination pair belonging to the subset. A network with a single component is *connected* whereas, if there is more than one component, a network is *disconnected*. For instance, isolated nodes count as disconnected components of the system. For a directed network, a component is defined as *weakly connected* if two nodes of the undirected representation of the component are connected by one or more paths. If there is a directed path in both directions between every pair of nodes, the component is defined as *strongly connected.*

For networks with finite size, we define the *largest connected component* (LCC) as the cluster with the maximal subset of nodes. If the size of the network is infinite, such as in the thermodynamic limit, the LCC is usually named a *giant connected component* (GCC).

In interdependent networks [129], two systems A and B are interconnected with links and the potentially functional clusters are identified by *mutually connected components*. If we indicate by $\mathcal{A}$ the set of nodes in network $G(A)$ and by $\mathcal{B}$ the corresponding set of nodes in network $G(B)$, they form a mutually connected component if:

- each pair of nodes in $\mathcal{A}$ is connected by a path consisting of nodes belonging to $\mathcal{A}$ and links of network $G(A)$;
- each pair of nodes in $\mathcal{B}$ is connected by a path consisting of nodes belonging to $\mathcal{B}$ and links of network $G(B)$ [68].

In multilayer networks, we can use the definition of path described in Section 3.1 to define a *multilayer connected component* as the subset of nodes

---

[5] Note that this notion is different from the one of groups or communities or modules, although the terminology might be sometimes misleading.

connected by a multilayer path [98]. This definition can be used to identify connected components from the aggregate representation of the multilayer network because it is sufficient that a pair of nodes is connected by a path to be part of the same component.

However, different and complementary information about the structure of a multilayer network can be provided by more restrictive definitions. For instance, the *largest intersection component* (LIC) is defined as the largest cluster in which nodes are connected across all layers simultaneously [39] and can be identified by intersecting the LCC of each layer separately. Another alternative is to aggregate the multilayer system with respect to the intersection of edges and then identify the LCC of the resulting network [230]. That definition has been recently used to better understand the emergence of continuous or abrupt percolation phase transitions even in systems of finite size.

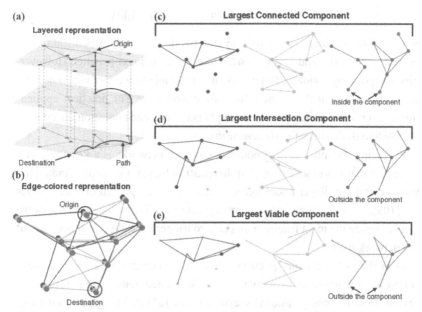

Figure 11 In many cases of interest, multilayer networks can be represented in different ways, for example, (a) in terms of interconnected layers – that is, a multiplex – or (b) by collapsing state nodes and explicitly encoding layers as colors of an edge-colored multigraph. To define components, it is important to identify origin and destination nodes and to consider paths connecting them. A multilayer connected component is identified by paths between state nodes. On the right-hand side of the figure, three distinct definitions of multilayer connected components are illustrated: largest connected component (c), largest intersection component (d), and largest viable component (e). See the text for details. Reproduced with permission from [93].

However, one more definition is possible and useful for applications. The *largest viable component* (LVC) is defined as the maximal subset of *viable* nodes and it consists of nodes that are connected by a path in each layer simultaneously [39]. As a result, all nodes in the LVC are essential to the function of the system and to define its structural core [27, 143, 269]. The more restrictive condition imposed by the LVC is responsible for a hybrid phase transition that leads to the discontinuous emergence of the giant viable cluster, at variance with ordinary percolation where a continuous phase transition is observed [39].

Figure 11 shows an illustration of multilayer connected components for both interconnected multiplex and edge-colored multigraph representations. The more restrictive the condition imposed, the smaller the size of the largest connected cluster: in particular, the size of the LVC is equal or smaller than the size of the LIC, which in turn is equal to or smaller than the size of the LCC.

3.3 Measuring Influence: Versatility

Information exchange with a multilayer network is strongly dependent on the organization of the underlying system into connected components. However, given a (possibly connected) network, it is plausible to wonder *which node(s)* are more important than others for information flow. For practical applications, several additional questions are plausible and each question encodes an operational definition of *node centrality*.

Revealing the most *central* nodes in complex networks is a key issue in a variety of real-world scenarios [30]. In epidemiology, for example, finding the most central multilayer nodes helps in identifying the pivotal disease spreaders [103], while in cascading failures, by detecting the most central nodes we can recognize the most fragile actors able to trigger the failure of other parts of the network [68].

Over the years, a wide spectrum of network centrality measures has been proposed for single-layer networks (for interested readers, we refer to a review from the computational social sciences [251]). The generalization of such descriptors to the realm of multilayer systems is not always immediate since nodes peripheral in one layer might be extremely central in another one [53, 97, 161], and crucial nodes for a given dynamics might not be central across all layers. If, for example, two distinct layers have only one node in common, any exchange of information between those layers will involve the passage through that common node independently from its centrality in each layer separately: hence, such a node will be highly central for the considered process. *Multilayer versatility* [97, 102] is a measure able to capture these features of a node, quantifying how important nodes are for the processes that characterize the definition of specific centrality descriptors – for example, in terms of information diffusion or flow through shortest paths – in a multilayer network.

In 2006, Borgatti and Everett [58] approached centrality from a graph-theoretic perspective, claiming that all centralities consider the involvement of a node in the walk structure of the network according to four features: Walk Type, Walk Property, Walk Position, and Summary Type. The *Walk Type* dimension concerns the kind of walks considered as well as the kind of constraints on such walks, for example, considering only shortest paths. The *Walk Property* dimension distinguishes between measures that consider the *volume* of walks a node is involved in (e.g., as in betweenness centrality) and measures that regard the *length* of those walks (e.g., as in closeness centrality). The *Walk Position* can be radial, where walks start in a given node (e.g., closeness), or medial, where walks pass through that given node (e.g., betweenness). Finally, the *Summary Type* dimension distinguishes measures according to the chosen summary statistic used to obtain a centrality score vector. Based on the possible combinations of these four features – that is, Walk Type, Walk Property, Walk Position, and Summary Type – different centrality measures can be obtained.

Besides identifying the most central nodes, the power of versatility lies also in predicting their role in some emblematic dynamical processes such as diffusion and congestion [102]. The importance of this aspect is highlighted, for example, when nodes of a multilayer network are ranked by their coverage [98] (see also Section 4.1) using the PageRank – see Section 3.3.7 for its definition – obtained from the multilayer model and its aggregated network. PageRank versatility, obtained from the multilayer network, is a better estimator of the evolution of such dynamics, outperforming its single-layer counterpart.

In the following, we present the most important centrality measures in the case of multilayer networks – that is, Multidegree, K-coreness, Eigenvector and Katz versatility, HITS versatility, Random walk occupation centrality, PageRank versatility, Random walk betweenness and closeness versatility, Betweenness versatility, and Interdependence centrality – while we refer the interested reader to other interesting measures that exploit multiplex features to identify a node's importance [228], recently validated to predict multiplex centrality in the rhesus macaque [43].

3.3.1 Multidegree

Multidegree centrality (k_i) is the simplest indicator of node importance at the local level, it is obtained by summing up all the links connected to node i across all layers [97]:

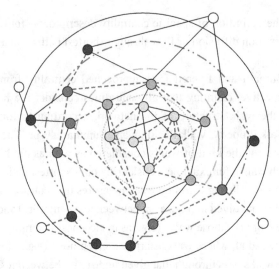

Figure 12 The (k_1, k_2)-core decomposition of a multilayer network characterized by two distinct types of edges. The first type is encoded by solid links, while the second type is encoded by dashed links. The concentric circles help identify the (k_1, k_2)-core decomposition: the cores from the most external circle to the innermost are the $(1,1)$-core, the $(1,2)$-core, the $(2,2)$-core, and the $(1,3)$-core, respectively. Figure from [27].

$$k_i = \sum_{\alpha,\beta=1}^{L} \sum_{j=1}^{N} M_{j\beta}^{i\alpha} = M_{j\beta}^{i\alpha} u_\alpha^\beta u^j. \tag{3.4}$$

In some cases – for example, when interlayer links are not explicitly considered – it can be more suitable to evaluate the degrees coming from individual layers:

$$k_i^\alpha = \sum_{j=1}^{N} A_j^i(\alpha), \tag{3.5}$$

where $A_j^i(\alpha)$ represents the adjacency tensor of layer α. Other definitions partially related to this one, for the case of noninterconnected multiplex networks, can be found in [34, 49].

3.3.2 K-coreness

In single layer network, the k-core of a graph is defined as a maximal connected subgraph in which each vertex has at least degree k within that subgraph. The ensemble of all k-cores of a graph represents the core decomposition of that specific graph [252]. To extend this concept to multilayer networks, we have to take into account the different types of edges, encoding different types of interactions encoded into layers. In this case, the k-core is defined as the largest subgraph in which each vertex has at least k_α edges of each type, $\alpha = 1, 2, ..., M$ where M is the total number of layers [27] (see Figure 12). For more detail on how to efficiently compute the complete core decomposition of a multilayer network, we refer to [124].

3.3.3 Eigenvector Versatility

Eigenvector centrality is a measure of influence of the single nodes in a network. A particular node has a high eigenvector centrality score when its neighbors have high scores themselves. [57]. In the monoplex case, the recursive character of this definition is untangled by the eigenvalue problem:

$$W\mathbf{x} = \lambda_1 \mathbf{x}, \tag{3.6}$$

where λ_1 is the largest eigenvalue of W and the element x_i represents the centrality score of the node i in the network described by the weight matrix W [97, 102]. If W is symmetric with positive entries, the Perron–Frobenius theorem grants the existence and uniqueness of this vector.

When complete knowledge of the intralayer connectivity is available, a natural extension of the definition of eigenvalue centrality for multilayer networks can be easily obtained as [97]:

$$\sum_{i,\alpha} M_{j\beta}^{i\alpha} \Theta_{i\alpha} = \lambda_1 \Theta_{j\beta}, \tag{3.7}$$

where λ_1 is the largest eigenvalue of M and Θ its corresponding eigentensor, whose values represent the centrality of each node in each layer. The problem of finding the eigenvector centrality consists in computing $\Theta_{j\beta} = \lambda_1^{-1} \sum_{i,\alpha} M_{j\beta}^{i\alpha} \Theta_{i\alpha}$, which represents the multilayer generalization of Bonacich's eigenvector centrality per node per layer [97, 102]. By summing up the scores of a node across all the layers, eigenvector versatility can be condensed across layers: $\theta_i = \sum_\alpha \Theta_{i\alpha}$.

Note that the summation across layers appears naturally through the tensorial formalism [97, 102]. Nevertheless, we can opt for other types of aggregation, based on heuristics specific to the nature of the centrality vector, to summarize the centrality measures computed for all layers into a unique descriptor. For other definitions, we refer the interested reader to [146, 257] and [274], the latter proposing tunable eigenvector-based centralities that can be applied to both temporal and multiplex networks.

3.3.4 Katz Versatility

In the case of directed networks, nodes with only outgoing edges have by definition eigenvalue centrality equal to 0. This may lead to meaningless results for the eigenvalue centrality, such as in the limiting case of a directed acyclic network, where all nodes have zero centrality [203]. This problem can be overcome by assigning to each node a minimum value of centrality so that all nodes are taken into account for measuring the influence between neighbors [158]. This can be realized [97] by redefining the eigenvalue problem as:

$$\sum_{i,\alpha}(aM_{j\beta}^{i\alpha}\Phi_{i\alpha} + 1) = \Phi_{j\beta} \tag{3.8}$$

with $a < 1/\lambda_1$, being λ_1 the largest eigenvalue of M. As for the eigenvector versatility, the overall Katz centrality of a node is the sum of centrality scores $\Phi_{j\beta}$ across layers [97, 102].

3.3.5 HITS Versatility

In single-layer networks, the Hyperlink-Induced Topic Search or HITS centrality (also known as *hub and authorities*) was originally introduced for web page rating according to their authority (e.g., their content) and their hub value (e.g., the value of their links to other web pages) [162]. Again in the case of directed graphs, it can be useful to recognize as important a node that is pointed to by important nodes or, alternatively, a node that points to important nodes. These two behaviors define two different roles a node can play in a directed network: hub and authorities. In the case of multilayer networks, how extensively a node plays the role of a hub or an authority is measured by the HITS versatility, which is defined by two eigenvalue problems [97, 102]:

$$\sum_{i,\alpha}(MM^{\dagger})_{j\beta}^{i\alpha}\Gamma_{i\alpha} = \lambda_1\Gamma_{j\beta} \tag{3.9}$$

$$\sum_{i,\alpha}(M^{\dagger}M)_{j\beta}^{i\alpha}\Upsilon_{i\alpha} = \lambda_1\Upsilon_{j\beta}, \tag{3.10}$$

where λ_1 is the largest eigenvalue (which is the same for the two problems), Γ represents the authority centrality, and Υ denotes the hub centrality.

3.3.6 Random Walk Occupation Centrality

A random walk on a network is a stochastic process where a path is defined, starting from a node of origin $X(0)$, with the node $X(t+1)$ chosen at random from among the neighbors of node $X(t)$. For a multilayer network, the probabilities of transition between pairs of nodes are represented as a tensor $T_{j\beta}^{i\alpha}$ [97] that are often taken as proportional to the edges' weights. If we let $p_{i\alpha}(t)$ be a time-dependent tensor giving the occupation probability of a given node in a given layer at time t, the random walk can be modeled as a Markov chain (note that we are using the Einstein convention):

$$p_{j\beta}(t+1) = T_{j\beta}^{i\alpha}p_{i\alpha}(t) \tag{3.11}$$

The steady state for this equation $\Pi_{i\alpha}$ can be obtained as the leading eigentensor in the eigenvalue problem:

$$T_{j\beta}^{i\alpha}\Pi_{i\alpha} = \lambda_1\Pi_{j\beta} \tag{3.12}$$

The probabilities $\Pi_{i\alpha}$ define the *random walk occupation centrality* [260], a measure that highlights which nodes might experience congestion due to insufficient outflow.

3.3.7 PageRank Versatility

For single-layer networks, PageRank [215] is a centrality measure originally developed for ranking web pages. It represents the occupation probability of random walkers on the network subjected to teleportation: at any time the walker can walk to a neighbor with a rate r and be teleported to any other node with rate $1 - r$. Its extension to a multilayer network can be based on different heuristics, in cases where the nature of the layer's coupling is unknown [146], or on the tensorial formulation [97, 102] if the interlayer connectivity is known. In this second case, the dynamics of the random walker are regulated by a transition tensor defined by:

$$R_{j\beta}^{i\alpha} = rT_{j\beta}^{i\alpha} + \frac{1-r}{NL}u_{j\beta}^{i\alpha}, \tag{3.13}$$

where $T_{j\beta}^{i\alpha}$ is the transition tensor of a classical random walk in the absence of teleportation and $u_{j\beta}^{i\alpha}$ is the rank-4 tensor with all components equal to 1. If we denote by $\Omega_{i\alpha}$ the eigentensor of $R_{j\beta}^{i\alpha}$, then the PageRank versatility ω_i of a node is obtained by summing across layers $\sum_{i,\alpha}\Omega_{i\alpha}$. In other words, the Page Rank centrality for multilayer networks is the steady-state solution of the master equation corresponding to the transition tensor $R_{j\beta}^{i\alpha}$.

It is important to remark that, for all nodes without outgoing edges, the transition tensor has to be redefined as $R_{j\beta}^{i\alpha} = \frac{1}{NL}u_{j\beta}^{i\alpha}$ to grant the correct normalization of the transition probabilities. We refer to [35, 146, 154] for existing variants on PageRank definitions in the context of multiplex networks.

3.3.8 Random Walk Betweenness Versatility

The random walk betweenness versatility [260] measures the importance of nodes in terms of the number of times random walk paths in the network pass by a given node. This quantity is suitable, for instance, for identifying the critical nodes in the random spreading of pieces of information, which are not necessarily following the shortest trajectories [121]. To analytically compute the random walk betweenness versatility, it is convenient to take advantage of the concept of absorbing random walks – that is, walks that will end when a given node d is reached. These walks are defined by the absorbing transition tensor on a particular node d:

$$
(T_{[d]})_{j\beta}^{i\sigma} = \begin{cases} 0 & j = d \\ T_{j\beta}^{i\sigma} & j \neq d. \end{cases} \tag{3.14}
$$

The average number of times a random walk (originating in node o in layer σ and destination d, independently of the layer) will pass by a node j in layer β, regardless of the time step, is given by [260]:

$$
(\tau_{[d]})_{j\beta}^{o\sigma} = \left[(\delta - T_{[d]})^{-1} \right]_{j\beta}^{o\alpha}, \tag{3.15}
$$

where $\delta_{j\beta}^{i\alpha} = \delta_j^i \delta_\beta^\alpha$ and δ is the Kronecker delta. The average over all possible starting layers σ and the aggregation of the walks that pass through j in the different layers are obtained by

$$
(\tau_{[d]})_j^o = \frac{1}{L} (\tau_{[d]})_{j\beta}^{o\sigma} u^\beta u_\sigma. \tag{3.16}
$$

Finally, the overall betweenness versatility is given by averaging over all possible origins and destinations:

$$
\tau_j = \frac{1}{N(N-1)} \sum_{d=1}^{N} (\tau_{[d]})_j^o u_o. \tag{3.17}
$$

3.3.9 Random Walk Closeness Versatility

Random walk closeness centrality [260] of a node i is defined as the inverse of the average number of steps required by a random walker, starting from any other node in the multilayer network, to reach i for the first time. As for the case of betweenness, by using the concept of absorbing random walks we can find that for walks originating in node o in layer σ the probability of visiting node j in layer β after t time steps is given by $(T_{[d]})_{j\beta}^{o\alpha}$ to the power of t. Considering the walk starting from node o in layer σ, each tensor encoding the mean first passage time to node d is given by [260]:

$$
(H_{[d]})^{o\sigma} = \left[(\delta - T_{[d]})^{-1} \right]_{j\beta}^{o\sigma} u^{j\beta}. \tag{3.18}
$$

The average mean first passage time $h_{[d]}$ to node d is obtained by averaging $(H_{[d]})^{o\sigma}$ over all possible starting nodes and layers as:

$$h_{[d]} = \frac{1}{(N-1)L} u_{o\sigma} (H_{[d]})^{o\sigma} + \frac{1}{N} \Pi_{[d]}^{-1}, \tag{3.19}$$

where $\delta_{j\beta}^{i\alpha} = \delta_j^i \delta_\beta^\alpha$ and $\Pi_{[d]}$ is the random walk occupation centrality. The last term $\frac{1}{N} \Pi_{[d]}^{-1}$ is included to account for the average return times, usually excluded when using absorbing random walks. Finally, the random walk closeness centrality of node d is simply obtained as the inverse of the distance $\xi_d = 1/h_{[d]}$.

3.3.10 Betweenness Centrality

If a metric distance can be defined in a multilayer topology, it is possible to extend the classical definitions of node betweenness centrality, edge betweenness centrality, and closeness centrality to obtain their multilayer counterparts [184, 196]. Given the existence of layers, it is possible to select a subset of layers Ω and consider only the paths belonging to that subset, defining the *cross betweenness centrality* [66] of the node i:

$$CBC(i, \alpha, \Omega) = \sum_{o \neq i, d \neq i, \beta \in \Omega, \beta \neq \beta'} \frac{\sigma_{o\beta}^{d\beta'}(i, \alpha)}{\sigma_{o\beta}^{d\beta'}}, \tag{3.20}$$

which counts the fraction of interlayer shortest paths, having their destination in Ω, that pass through node i of layer α. Taking advantage of this definition, the multiplex betweenness centrality (BC) can be decomposed into the following contributions:

$$BC(i, \alpha) = CBC(i, \alpha, \Omega) + CBC(i.\alpha, \bar{\Omega}) + IBC(i, \alpha), \tag{3.21}$$

where $\bar{\Omega}$ indicates the layers not belonging to Ω and $IBC(i, \alpha)$ is called *internal betweenness centrality* and represents the contribution at the BC of the paths that never leave the layer α:

$$IBC(i, \alpha) = \sum_{o \neq i, d \neq i, \alpha = \beta = \beta'} \frac{\sigma_{o\beta}^{d\beta'}(i, \alpha)}{\sigma_{o\beta}^{d\beta'}}. \tag{3.22}$$

Other definitions are also available and define betweenness versatility as another natural extension of the monoplex betweenness centrality [259, 261].

3.3.11 Interdependence Centrality

In general, the presence of more than one layer within a system increases the number of available paths with respect to the case where only one single layer is present. The richness of multilayer paths allows for the possibility that single-layer shortest paths are longer than multilayer ones. To quantify the value added by multiplexity to potential communication routes, it is possible to define the reachability of each physical node i in terms of interdependence centrality [206], defined as:

$$\lambda_i = \frac{1}{N-1} \sum_{j \neq i} \frac{\psi_{ij}}{\sigma_{ij}}, \tag{3.23}$$

where ψ_{ij} is the number of shortest paths between nodes i to j that pass by more than one layer and σ_{ij} is the total number of shortest paths from i to j. In both cases, the number of shortest paths can be greater than 1 if there are multiple paths having the same, minimal length. Note that these quantities are not tensors or matrices: they can be better understood in terms of matrix entries, since they encode pairwise information about physical nodes. The global interdependence is computed by summing up over the interdependence centrality of all nodes: $\lambda = \langle \lambda_i \rangle$.

In some systems, such as human transportation, the flows between different pairs of nodes might be strongly dis-homogeneous. In such cases, the global interdependence can be redefined to take into account the real loads of the network by weighting the aforementioned sums by T_{ij}, which is an origin-destination matrix whose entries are normalized such that $\sum_{i,j} T_{ij} = 1$. The weighted interdependence, also known as *coupling* [196], is therefore given by

$$\lambda' = \sum_{i,j \neq i} T_{ij} \frac{\psi_{ij}}{\sigma_{ij}}. \tag{3.24}$$

3.4 Mesoscale Organization

3.4.1 Triadic Closures

The clustering coefficient, a measure of the cohesiveness of triads, is an important topological property of a node and a whole network. Nevertheless, it can also be defined as the number of 3-cycles that start and end at the same node [82], relating this network descriptor to how information flows and localizes. As exemplified in Figure 13 for an interconnected multiplex network, a 3-cycle can go through different layers, but is characterized only by 3 intralayer steps. These type of walks – in which after or before each intralayer step the walker can choose between continuing on the same layer or changing to some other adjacent layer – can be explicitly decomposed into combinations of intra- and interlayer steps by using adequately defined tensors:

$$\mathcal{A}_{i\alpha}^{j\beta} = \begin{cases} M_{i\alpha}^{j\beta} & \alpha = \beta \\ 0 & \alpha \neq \beta, \end{cases} \tag{3.25}$$

encoding intralayer steps, and

$$C_{i\alpha}^{j\beta} = \begin{cases} M_{i\alpha}^{j\beta} & i = j \\ 0 & i \neq j, \end{cases} \tag{3.26}$$

encoding interlayer steps. Since changing layers is optional, we represent choice of layers as

$$\hat{C}_{i\alpha}^{j\beta} = bI_{\alpha}^{\beta} + gC_{i\alpha}^{j\beta}, \tag{3.27}$$

where I is the identity matrix, b represents the weight of staying in the layer i, and g denotes the weight of stepping from layer α to layer β.

With this definition, two types of cycles can be defined, and therefore two different number of cycles starting in i can be obtained. In the first case, in which two consecutive interlayer steps are forbidden, this number is $t_{(w),i\alpha} = 2\left((\mathcal{A}\hat{C})^3\right)_{i\alpha}^{i\alpha}$, corresponding to the examples shown in Figure 13), while in the second case, if those steps are permitted, this number is $t_{(sw),i\alpha} = \left((\hat{C}\mathcal{A}\hat{C})^3\right)_{i\alpha}^{i\alpha}$.

To compute the multiplex clustering, a normalization factor $d_{(*),i\alpha}$ is also required, where with the label $(*)$ we indicate both types of cycles, (w) and (sw). The corresponding definitions for both factors d can be obtained from the t numbers by replacing the tensor $\mathcal{A}$ in the second intralayer step with the tensor:

$$F_{i\alpha}^{j\beta} = \begin{cases} 1 & i \neq j \text{ and } \alpha = \beta. \\ 0 & \text{elsewhere} \end{cases} \tag{3.28}$$

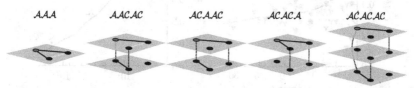

$$\mathcal{A}\mathcal{A}\mathcal{A} \qquad \mathcal{A}\mathcal{A}C\mathcal{A}C \qquad \mathcal{A}C\mathcal{A}\mathcal{A}C \qquad \mathcal{A}C\mathcal{A}C\mathcal{A} \qquad \mathcal{A}C\mathcal{A}C\mathcal{A}C$$

Figure 13 An elementary cycle on a multilayer network can include edges from up to three different layers, identifying a triad of physical nodes. The starting point of the cycle is in the leftmost node of the upper layer, intralayer edges are dotted curves, and intralayer edges are solid lines. To simplify the reading of the cycle, the second intralayer step is represented by a white line. Above each cycle is indicated the tensorial form of the cycle in terms of the tensors $\mathcal{A}$ and C described in the text. Figure from [82].

For example, for the $_{(w)}$ type of cycles we have $d_{(w),i\alpha} = 2\mathcal{A}\hat{C}F\hat{C}\mathcal{A}\hat{C}$. The local multiplex clustering coefficient for the state node (i, α) can be then calculated as:

$$C_{(*),i\alpha} = \frac{t_{(*),i\alpha}}{d_{(*),i\alpha}}, \tag{3.29}$$

while the global multiplex clustering coefficient, the scalar value representative of triadic closure in the whole system, is given by

$$C_{(*)} = \frac{\sum_{i\alpha} t_{(*),i\alpha}}{\sum_{i\alpha} d_{(*),i\alpha}}. \tag{3.30}$$

We can decompose Equation (3.30) into its contributions coming from cycles traversing a different number m of layers as

$$C_{(*)}^{(m)} = \frac{\sum_{i\alpha} t_{(*),i\alpha}^{(m)}}{\sum_{i\alpha} d_{(*),i\alpha}^{(m)}}, \tag{3.31}$$

where $t_{(*),i\alpha}^{(m)}$ and $d_{(*),i\alpha}^{(m)}$ are restricted to cycles involving exactly m layers, with $m = 1, 2, 3$. This perspective can be useful to appreciate the differences observed for cycles of a different nature, evident in Figure 14, where local clustering coefficients are drawn as a function of the nodes' degrees. In several empirical multiplex systems [82], it has been observed that $C < C^{(1)}$ and

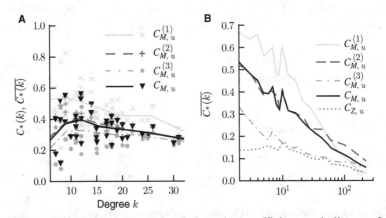

Figure 14 Relation between local clustering coefficient, an indicator of triadic closure, and node's unweighted degrees for m-layers cycles $C_{M,u}^{(m)}$ as defined in Equation (3.31), and for all cycles, but considering the multilayer network with its unweighted $C_{M,u}$ or weighted $C_{Z,u}$ representation. Two empirical systems are considered: (a) Kapferer tailor-shop social network; (b) the airport network obtained from openflights.org. The curves interpolate the local clustering coefficient for each type of cycle. Figure from [82].

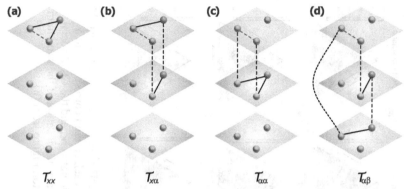

Figure 15 Four possible types of triadic closures on a multilayer network that can be used for predicting the existence of missing links: (a) $\mathcal{T}_{xx}$, if the two links belong to the same layer x where the cycle is closed; (b) $\mathcal{T}_{x\alpha}$, if one link belongs to x and another to a different layer α; (c) $\mathcal{T}_{\alpha\alpha}$ if both existing links belong to the same layer, different from x; (d) $\mathcal{T}_{\alpha\beta}$ if the two existing links belong to two separate layers, both different from x. Figure from [8].

$C^{(1)} > C^{(2)} > C^{(3)}$ as a general pattern that provides a kind of hierarchy of triadic closure across layers.

It is worth noticing that triadic closure can also be used to predict future or missing links on a multilayer network, as recently shown in [8]. As can be seen from Figure 15, depending on the location of the links, there are four possible triadic relations that allow us to predict a missing link by closing a cycle of two links, u on layer α and v on layer β, with a third link on layer x. This observation allows for a multilayer generalization of the Adamic–Adar method [4], which predicts links on the basis of a score based on the number of common neighbors weighted by their degree.

In fact, in multilayer networks, the neighbors of a node can belong to different layers. An appropriate "Multilayer Adamic–Adar" (MAA) score counts the common neighbors closing the triads of each of the aforementioned types, weighting each contribution by the logarithm of the degree as usually done to calculate the Adamic–Adar score:

$$MAA(u,v) = \sum_{\alpha,\beta} \sum_{w \in \mathcal{T}_{\alpha\beta}} \frac{\eta_{x\alpha}\eta_{x\beta}}{\sqrt{\langle k \rangle_\alpha \langle k \rangle_\beta}} \cdot \frac{1}{\sqrt{\ln(k_w^\alpha)\ln(k_w^\beta)}} \tag{3.32}$$

where $\langle k \rangle_\alpha$ indicates the average degree of nodes in layer α, k_w^α the degree of node w in layer α, and $\eta_{x\alpha}$ are free coefficients allowing to control the relative weight of each type of triadic closure in the link's total score.

Figure 16 shows the AUC (panel a) and the precision (panel b) of the MAA for three empirical systems as a function of the coefficients $\eta_{x\alpha}$. The results

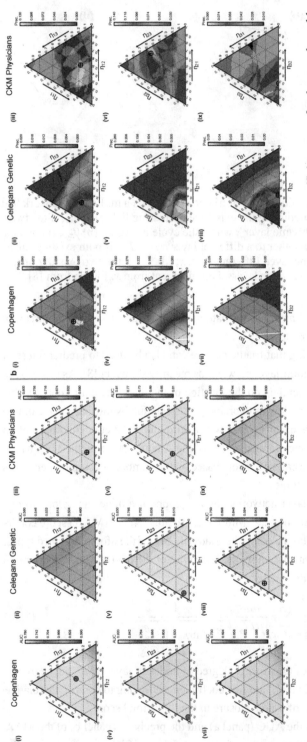

Figure 16 How areas under the curve (AUC) (a) and precision (b) change for different values of the coefficients $\eta_{\beta\alpha}$, for three real-world multiplex networks. For simplicity, in each network only three layers are considered. In both panels, each column corresponds to a different data set, from left to right: a social network of students in Copenhagen, the *C. elegans* genetic network, and the CKM social network of physicians. Each row corresponds to a prediction in a different layer. The cross marker indicates the location of the maximum value for each plot, corresponding to the combination of coefficients optimizing the indicator. Figure from [8].

show that single-layer link prediction can be largely improved by exploiting information encoded in other layers.

3.4.2 Communities and Modules

Complex systems are characterized by the mesoscale organization of their units into groups, also known as *modules* or *communities* [118, 120, 202]. However, the definition of what a group exactly corresponds to is an open problem. Here, we will briefly describe the major advances in this direction, considering four methods widely used in the literature, namely multilayer modularity maximization, multilayer tensor factorization, multilayer Bayesian inference, and multilayer description length minimization through the map equation. A variety of methods is available, including the analysis of intermittent communities in time-varying networks [23], but they are beyond the scope of this work and deserve a dedicated review [148, 149].

Multilayer Modularity Maximization

The simplest – and also one of the most popular ones – definition concerns the density of links within a group with respect to intergroup density of links: for a module, we expect units to be mostly connected with other units inside the module and poorly connected outside. This approach is based on the calculation of a function named *modularity*, which, for a given partitioning of the units into groups, quantifies the deviation of the number of links from what is expected by chance according to the corresponding configuration model [201]. This function is therefore calculated for all possible partitions, and the partition with maximum modularity is chosen as the most representative mesoscale organization of the underlying network, providing a hint about its structural and functional organization. For this reason, identifying such an organization is of paramount importance for applications in many disciplines, from social sciences to biology [133, 144].

The definition of the multilayer generalization of modularity was given by [198] a decade ago. It has become a standard for applications [12, 72]. It assumes that the system can be represented by a three-dimensional tensor, encoding within-layer connectivity as for edge-colored multigraphs and time-varying networks (see Figure 17) and by another three-dimensional tensor encoding interlayer connectivity.[6] To allow for a multi-resolution analysis, a resolution parameter γ is introduced and interlayer connectivity is weighted

---

[6] Where each snapshot is encoded by a layer.

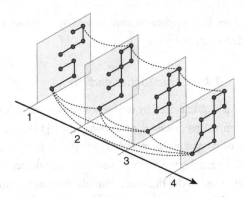

Figure 17 Illustration of a *multislice* network representing, for instance, a time-varying network where each temporal snapshot is encoded by a layer and interlayer connectivity encodes temporal relationships among state nodes. Figure from [198].

by another tunable parameter ω.[7] Using the tensorial notation, considering the more general systems that can be represented by the multilayer adjacency tensor $M_{j\beta}^{i\alpha}$, and indicating by $P_{j\beta}^{i\alpha}$ the tensor encoding the connectivity values obtained from a null model (e.g., the configuration model), we can define the rank-4 tensor

$$B_{j\beta}^{i\alpha} = M_{j\beta}^{i\alpha} - P_{j\beta}^{i\alpha}, \tag{3.33}$$

which is then used to define the modularity function

$$Q \propto S_{i\alpha}^{a} B_{j\beta}^{i\alpha} S_{a}^{j\beta}, \tag{3.34}$$

to be maximized while varying the membership tensor $S_{a}^{i\alpha}$, with entries equal to 1 if node i in layer α belongs to the community a, and 0 otherwise [97]. This approach can be suitably used to identify groups in time-varying systems, like the US Senate (Figure 18). We refer to [41] for an enhanced version of this method and to the recent work demonstrating that, under certain conditions, modularity maximization corresponds to maximizing the posterior probability of community assignments under suitably chosen stochastic block models [216]. For further information about the latter, see also later in the current subsection on multilayer Bayesian inference later in this text.

---

[7] We refer the interested reader to [18, 119, 273] for more information about the resolution and super-resolution problems in community detection. Note that recent studies have shown that by aggregating layers by summation, for instance, it is also possible to enhance the detectability of community structures [272].

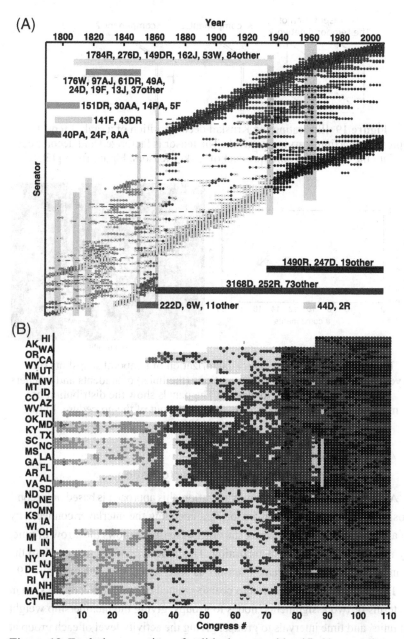

Figure 18 Evolution over time of political groups, identified by multilayer modularity maximization, in the US Senate across more than two centuries (110 layers encoding the number of two-year Congresses from 1789 to 2008). Intralayer links are given by roll call vote similarities and interlayer connectivity is weighted by $\omega = 0.5$. (A) Results from senators' assignments. (B) Results from state affiliations. Note that less sophisticated analytical techniques, applied, for instance, on aggregated representation of this system, would not identify the same patterns. Analysis and figure from [198].

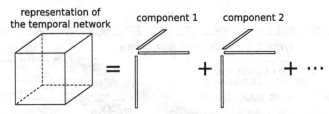

Figure 19 Illustrating the Kruskal decomposition of a rank-3 tensor representing a time-varying network. The tensor is factorized and decomposed into the sum of the outer product of rank-1 tensors. Figure from [130].

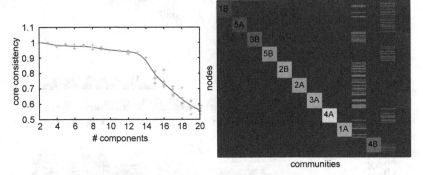

Figure 20 Nonnegative tensor factorization of temporal snapshots (the layers) describing close-range interactions (the links) of students and teachers (the nodes), divided in to classes. The panels show the distribution of membership weights encoded by the first output of the procedure. Figure from [130].

Multilayer Tensor Factorization

At variance with modularity maximization, this approach is based on decomposing the three-dimensional tensor $T_{ij\alpha}$, encoding the interlayer connectivity of a time-varying network where the order of tensors follows the arrow of time, by means of nonnegative factorization. This procedure maps a tensor into the sum of outer products of rank-1 tensors as schematically shown in Figure 19.

The result of this procedure is a coarse-grained representation of the system in terms of two assignments: nodes to groups, giving the membership weight of units, and time intervals to groups, giving the activity level of each group at different temporal snapshots. When applied to a contact network, for instance, the result of this approach is shown in Figure 20.

Multilayer Bayesian Inference

A different approach is to consider the problem of recovering the best partition of the nodes, where we assume that there is an underlying model encoding some mechanisms at work to generate the observed system. This class of methods, originally developed for analysis of social networks [147, 213, 256], assumes that groups can be encoded by blocks: probability to have links within and between blocks, as well as the number of blocks, are the parameters of the model to be calculated. This model is known as the stochastic block model (SBM), and it finds extensive applications in machine learning [5, 13, 134, 229]. More recently, a variation based on Bayesian inference and statistical physics has been proposed for applications to multilayer networks [221, 285]. A thorough discussion of the mathematical framework behind this powerful approach is beyond the scope of this section; therefore we refer the interested reader to an accurate and recent work fully devoted to this purpose [223].

Here, we outline the basic idea behind this method [221], which is schematically summarized in Figure 21 for the case of a monoplex. The input is considered an edge-colored multigraph or a time-varying network that can be represented by an array of matrices, similar to the ones we have described in the previous sections. Let us indicate this object with M_{ij}^{α}, and its aggregation obtained by entry-wise summation across layers as $A_{ij} = M_{ij}^{\alpha} u_{\alpha}$. Let Θ indicate the set of all parameters of the model to fit. The Bayes theorem in this case reads

$$P(\Theta|M_{ij}^{\alpha}) = \frac{P(M_{ij}^{\alpha}|\Theta)P(\Theta)}{P(M_{ij}^{\alpha})}, \tag{3.35}$$

where $P(\Theta)$ is the prior probability on the parameters, $P(M_{ij}^{\alpha})$ is a normalization factor, and $P(M_{ij}^{\alpha}|\Theta)$ is the likelihood of observing the multiplex system encoded by $M_{ij}^{\alpha} u_{\alpha}$ given the parameters. Here, $P(\Theta|M_{ij}^{\alpha})$ is the posterior likelihood: the higher its value the more likely our model describes the data. By noticing that:

- $P(\Theta) = e^{-\mathcal{L}(\Theta)}$, where $\mathcal{L}(\Theta)$ is the microcanonical entropy of the parameter ensemble;
- $P(M_{ij}^{\alpha}|\Theta) = e^{-S(M_{ij}^{\alpha})}$, where $S(M_{ij}^{\alpha})$ is the microcanonical entropy of the model ensemble;
- the term $P(M_{ij}^{\alpha})$ can be neglected since it does not depend on parameters and only acts as a constant;

it is straightforward to show that minimizing the function[8]

$$\Sigma = S(M_{ij}^{\alpha}) + \mathcal{L}(\Theta), \tag{3.36}$$

---

[8] Hint: take the natural logarithm of both sides of the Bayes formula, discard the constant term, and solve by Σ.

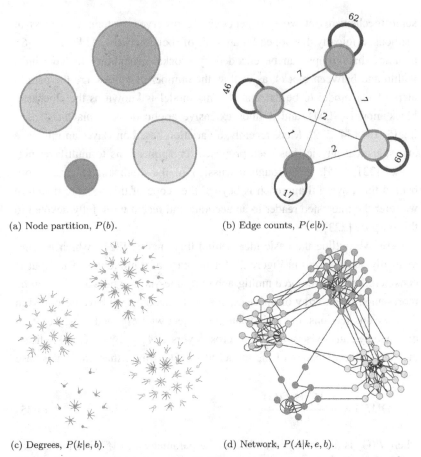

(a) Node partition, $P(b)$.   (b) Edge counts, $P(e|b)$.

(c) Degrees, $P(k|e,b)$.   (d) Network, $P(A|k,e,b)$.

Figure 21 Summarizing a nonparametric generative process in the case of a monoplex network, according to degree-corrected stochastic block modeling (DCSBM): (a) sampling the partition of units; (b) counting edges between and within groups; (c) imposing the observed degree sequence; (d) sampling the overall network by accounting for (a)–(c). Figure from [222].

known as the *description length*, is equivalent to maximizing the posterior probability. Note that Σ has a very nice interpretation in terms of information theory: it is the number of bits required to describe the data summed to the number of bits required to describe the model.

The explicit definition of the prior probability $P(\Theta)$ is often a subtle issue involving the choice of a generative process for the parameters. This is a problem-specific task that depends on the a priori knowledge and assumptions about the data. In a more general setting, and in those cases in which the prior knowledge is missing, it is desirable to choose *uninformative* priors. In [221], the author proposes a nested uninformative prior that minimizes the influence on the posterior.

We can choose different models to generate the observed network and, specifically, plausible options are to either generate layers conditional to the aggregate representation of the system or to generate each layer independently from the others (Figure 22), as well as other generative models – such as the Multilayer degree-corrected stochastic block modeling (M-DCSBM) – specifically designed for multilayer systems [42]. The reader could be concerned about how to choose the best model: in fact, the framework also allows for model selection and we should take the model that minimizes the description length. When we need a more refined approach, where a model or its alternative should be chosen, nonparametrically, with some degree of confidence, it is possible to calculate the posterior odds ratio as

$$\Lambda = \frac{P(\Theta_a|M_{ij}^\alpha, \mathcal{H}_a)P(\mathcal{H}_a)}{P(\Theta_b|M_{ij}^\alpha, \mathcal{H}_b)P(\mathcal{H}_b)} = e^{-\Delta\Sigma}\frac{P(\mathcal{H}_a)}{P(\mathcal{H}_b)}, \tag{3.37}$$

where each model class is encoded into the hypothesis $\mathcal{H}_p$ ($p = a, b$), $P(\mathcal{H}_p)$ is the prior belief for that hypothesis, and $\Delta\Sigma = \Sigma_a - \Sigma_b$ is the difference between the description length of each hypothesis. The operational prescription is that if $\Lambda < 1$, then $\mathcal{H}_b$ should be preferred to $\mathcal{H}_a$, whereas when $\Lambda \approx 1$, both models are equally plausible. More generally, if $\Lambda \in [1/3, 1]$, the difference between the two hypotheses can be considered negligible. An application of this procedure is shown in Figure 23 in the case of an empirical social network.

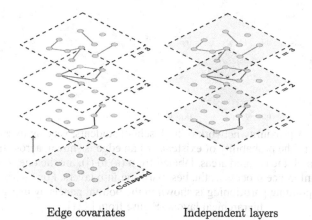

Edge covariates   Independent layers

Figure 22 Generative models of multilayer networks. Left: generate the aggregated (collapsed) representation of the system and then create layers whose aggregation leads to the same representation. Right: each layer is generated independently from the others. Figure from [221].

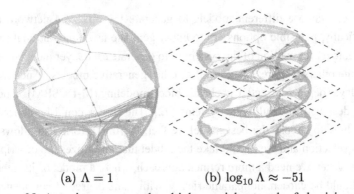

(a) $\Lambda = 1$ (b) $\log_{10} \Lambda \approx -51$

Figure 23 A noninterconnected multiplex social network of physicians is reproduced by means of two distinct generative models. The layout is specially designed to highlight the presence of hierarchies, indicated by square markers and links, of nodes – the circular dots arranged on a circle, linked by observed edges drawn by means of bundling to better put in evidence intergroup connections. (a) the inferred DCSBM is directly applied to the aggregated network; (b) inference is performed while accounting for edge covariates by assigning each layer to a type of social interaction. Below each panel, the value of the posterior odds ratio is shown. Figure from [221].

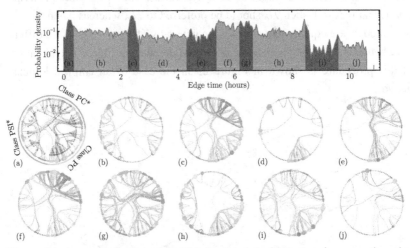

Figure 24 A proximity network of high school students, varying over time, is considered. The probability of existence of an edge is shown, across time, in the top panel: the colored areas, labeled from (a) to (j), are the ones obtained from the inference process as the best partition into layers. For each area, the corresponding partitioning is shown in the bottom panels by using the hierarchical layout. Figure from [221].

Finally, this procedure can be used to identify the best division of time-varying edges into layers, as shown in the case of a real-world contact network in Figure 24.

We refer the interested reader to a more recent work [217], where an SBM-like model – not relying on sampling of edges in different layers independently – is proposed to account for edge correlations and to define a new measure of layer–layer correlation that incorporates similarity between connectivity patterns in different layers. It is used for link prediction in multilayer networks.

Multilayer Description Length Minimization through the Map Equation

Instead of relying only on topological information – such as the abundance of edges or the presence of blocks – to identify groups, we can observe how information flow is trapped within modules, the rationale being that information is more likely to flow within a group than between groups [169, 226]. Another approach might be to map the potential regularities of the flow into strings and to look for the partitioning of the system that minimizes the corresponding description length [113, 237, 238], as schematically described in Figure 25.

The generalization to multilayer networks of this method, known as Multiplex InfoMap, was introduced in [99] and, more recently, it was better formalized under the perspective of higher-order modeling [112, 170, 239, 244], as schematically illustrated in Figure 26.

Multiplex InfoMap is based on the same principles of its single-layer counterpart, with some important differences. First, the monoplex InfoMap is based on Markovian dynamics, whereas this feature is kept only for within-layer dynamics. Second, coding only captures the visits to physical nodes, not the ones to state nodes, an effective non-Markovian dynamics across layers. The remainder of the method is the same: compressing persistent multilayer trajectories to identify modules according to the map equation[9]

$$\Sigma(\mathcal{M}) = q_{in}S(Q) + \sum_{\ell=1}^{m} p_\ell S(\mathcal{P}_\ell),$$ (3.38)

which encodes information flows within and across layers, as shown in Figure 27. This equation deserves a careful explanation of its terms.

Here, $\mathcal{M}$ is a partitioning of the system – that is, a model for the observed modules – and Σ is its description length, in bits. Indicating with $q_{\ell,in}$ and $q_{\ell,out}$ the transition rates at which a random walker enters and exits, respectively, a module ℓ, with $q_{in} = \sum_\ell q_{\ell,in}$ the sum of those rates and with $S(Q)$ the information entropy of the normalized probability distribution of the transition rates

---

[9] The curious reader might wonder about the governing equations of the corresponding random walk: this will be discussed in the section devoted to dynamics and, specifically, in Section 4.1.

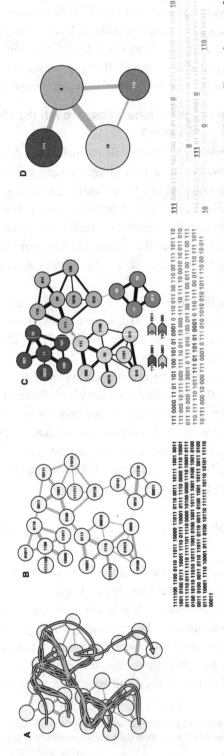

Figure 25 Detection of modules based on the Infomap algorithm. (A) The solid line represents a trajectory of random walks on the top of a classical network. (B) By encoding each node with a code word – for example, using a Huffman scheme – we can efficiently describe the trajectories in terms of strings by appending the code word of each node visited by the walker. The trajectory shown in (A) is encoded into a string of 314 bits. (C) Given a partitioning of the system, the persistence of the trajectory in a certain module is encoded by a certain number of bits: the lower this number the more persistent the trajectory in that module. This regularity can be used to coarse-grain the network into functional modules as in (D). Figure from [238].

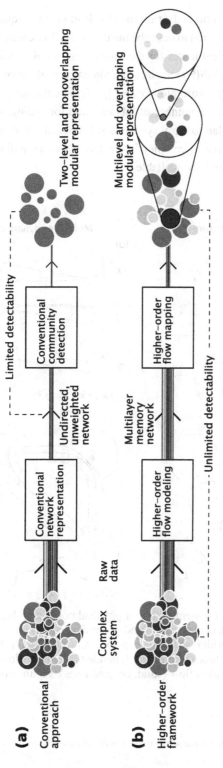

Figure 26 A complex system is analyzed using a conventional approach (a) and the higher-order framework (b) provided by the map equation. The latter allows us to account for the richness of the data, such as weights and directionality, without being affected by limited detectability (e.g., due to the resolution problem mentioned for modularity maximization) and, remarkably, is suitable to detect multilevel and overlapping modular information within the same elegant mathematical framework. Figure from [112].

($Q = \{q_{\ell,in}/q_{in}\}$), the first term in the right-hand side of the map equation captures the number of bits required to describe the dynamics. Indicating with $p_{l\in\ell}$ the visit rates of the physical nodes for the module codebook ℓ, with p_ℓ the sum of those rates, and with $\mathcal{S}(\mathcal{P}_\ell)$ the information entropy of the corresponding normalized probability distribution ($\mathcal{P}_\ell = \{p_{l\in\ell}/p_\ell\}$), the second term of the map equation captures the number of bits required for coding. An application of Multiplex InfoMap to a toy system is schematically summarized in Figure 27. The interested reader can play interactively with an online version of this method, which is publicly available.[10]

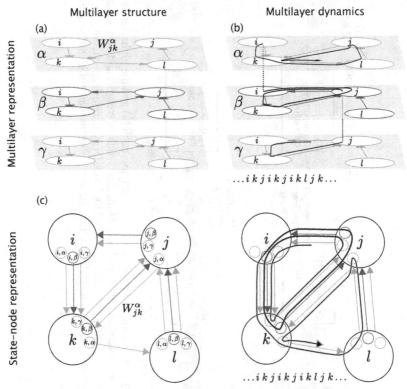

Figure 27 Toy multilayer system with four physical nodes and three layers, represented in terms of structure (a)–(c) and dynamics (b)–(d) with a layered perspective (a)–(b) and with respect to state-node representation (c)–(d), the latter highlighting the existence of memory (encoded by small colored nodes), which alters the information flow (solid, colored lines). Figure from [99].

[10] www.mapequation.org/apps/sparse-memory-network/index.html.

3.4.3 Integrated and Segregated Systems

To understand how a network operates, it is essential to understand if information flows in such a way that, from a functional perspective, the system is either integrated – that is, operating like a whole – or segregated – that is, operating like independent modules or groups of units (see Figure 28).

How to correctly study integration and segregation of a complex network is a still debated topic, and several proxies have been proposed across a broad spectrum of disciplines. The analysis stems almost independently from sociology and neuroscience, in both cases within the frameworks of classical single-layer networks [3, 47, 76, 127, 171, 172, 183, 240, 302].

In multilayer systems, where multiple relationships coexist simultaneously, the evaluation of integration must be accounted for by more complex topological models. In fact, it is critical to choose which kind of paths are to be evaluated – for example, by adding a cost to interlayer links used in paths that allow us to switch between two or more layers. The literature on this topic is rather poor and an agreement on the most suitable methodology to adopt is far from being reached. In the following, we describe two recent attempts in this direction.

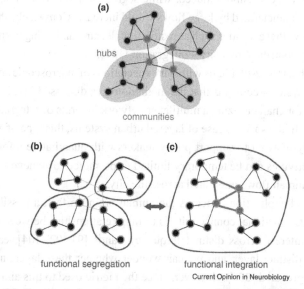

Figure 28 Segregated communities and integrating hubs. (a) Network with four communities interconnected by four central hubs. (b) Functional segregation due to community structure is estimated by within-community coupling. (c) Functional integration favors information flow across the whole network through a subset of interconnecting hubs and is measured by the hubs or the paths crossing the network. Figure from [266].

The first application comes from the analysis of socio-technical systems, and in particular movement flows. One important approach in this direction is communicability [6], for which an explicit multilayer definition is available [116]. However, the simultaneous analysis of integration and segregation can provide a more comprehensive perspective of a city's structural form and functional behavior. In a city, distinct layers can represent different activities (e.g., leisure, eating, shopping, sport, etc.; see Figure 29) that can be performed in the same area: two areas are connected within or across layers if there is a human flow between them in either direction. The analysis of empirical flow networks of this type measured from 10 megacities around the world would reveal, for instance, which activity was most influential for the urban system, from an integration–segregation perspective [127]. Using modularity as a proxy for integration and normalized global efficiency as a proxy for integration allowed scholars to unravel the functional organization of each megacity and to discover the presence of distinct "cities within a city." By comparing the deviations measured after aggregating all layers into a single-layer network and after aggregating while excluding one layer at a time, it can be shown that "transportation" is the activity whose disruptions would remarkably hinder the urban function: in fact, the integration granted by long-range connections would be sensibly reduced, while segregation due to the interrupted connections characterized by large flows would increase. Conversely, the same analysis shows that activities such as restoration, leisure, and going to shopping malls poorly contribute to a city's integration [127].

The fact that this result aligns with our expectations of macroscopic and easy-to-interpret systems confirms that the methodologies discussed in this section are suitable for characterizing a multilayer network in terms of integration and segregation. In the specific case of layered urban systems, this type of analysis is potentially relevant to support policy makers with quantitative information on which activities can be temporary limited or promoted to achieve a desired amount of human flows integrated across the city.

The second application concerns the human brain. In fact, a possible way to analyze its functional connectivity is to map relationships between distinct regions of interest across distinct frequency bands [91, 92, 104], each one encoding a distinct layer, and we can wonder whether these layers are integrated or segregated [276]. However, since the metric used to this aim is edge overlap, according to the aforementioned prescriptions this application is not entirely compatible with the study of integration and segregation and would be better understood as an analysis of layer–layer correlations. It will be interesting to see more studies on this topic in the future –, for instance, based on multilayer modularity and multilayer generalization of global efficiency.

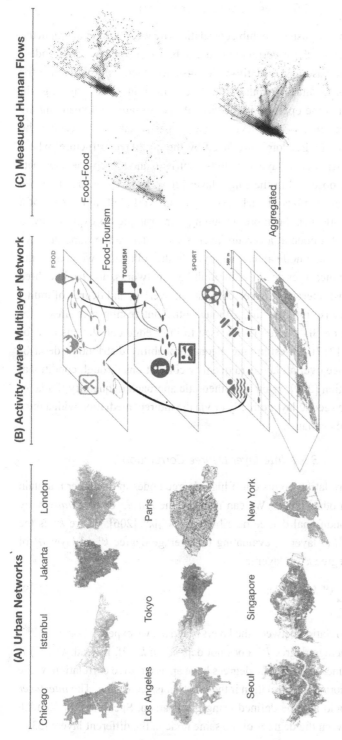

Figure 29 Multilayer modeling of structure and function of urban systems. (A) Urban backbone of 10 megacities described by their street networks. (B) Urban functional networks described by the flows between cells of size 500m×500m. The Foursquare data used for this analysis allow us to disentangle mobility flows into different layers where subsequent check-ins have been made between activities of the same type (e.g., leisure, eating, shopping, sport, etc.). (C) Measured flows for New York City represented as the weighted edges, which may be very different across distinct activity layers. Figure from [127].

3.5 Layer–Layer Correlations

Single-layer networks usually exhibit correlations between entities acting in the system. Specifically, these correlations occur between some specific nodes' properties. In the case of assortativity, for instance, there is a preference for network's nodes to attach to others that are similar in some way (e.g., in degree), while it is the opposite in the case of disassortativity. Expanding the concept of *correlation* to multilayer networks has to take into account the increased degrees of freedom introduced by the multilayer structure, where nodes' involvement across layers exhibits nontrivial and more variegated patterns than those observed in the single layer [205]. In fact, we can study the degree–degree correlations of each layer in the network, but, in the case of a multilayer network, it is far more intriguing, for example, to explore how a given property of a node at a certain layer is correlated to the same or other properties of the same node at another layer. In the following, we present the main correlation measures designed for multilayer networks, divided into three main subsections: interlayer degree correlation, overlap, and degree of multiplexity and pairwise multiplexity, although other interesting measures – for example, based on SBM-like models used to infer edge correlations and for link prediction [217], on estimating a joint probability distribution describing edge existence over all layers to quantify correlations through conditional mutual information [299], or on a set-theoretic approach to quantify if a layer correlates with a second layer directly or via the indirect mediation with a third layer [168] – are available.

3.5.1 Interlayer Degree Correlation

This kind of correlation points out if high-degree nodes in one layer maintain this property in other layers. We can measure the *interlayer assortativity* by studying the conditional degree distribution $P(k^{\alpha'}|k^{\alpha})$[206], where k_i^{α} is the degree of node i at layer α, evaluating the average degree $\bar{k}^{\alpha'}$ at layer α' of nodes having degree k^{α} at layer α:

$$\bar{k}^{\alpha'}(k^{\alpha}) = \sum_{k_{\alpha'}} k^{\alpha'} P(k^{\alpha'}|k^{\alpha}); \tag{3.39}$$

if there is no correlation between the layers α and α', we expect $\bar{k}^{\alpha'}(k^{\alpha}) = \langle k^{\alpha'} \rangle$ – that is, the average degree $\bar{k}^{\alpha'}$ does not depend on k^{α}. If, instead, $\bar{k}^{\alpha'}(k^{\alpha})$ is an increasing function in k^{α}, the degrees have an assortative correlation, while they have a dissortative correlation if the function is decreasing. The *interlayer assortativity* could also be defined using the Pearson, Spearman, or Kendall correlation between the degrees of the same node in the different layers [53].

3.5.2 Overlap and Degree of Multiplexity

In this case, correlation is evaluated in terms of node connectivity patterns in different layers. Specifically, we refer to the correlations among the links of the different intralayer networks. In fact, the internal connectivity in different layers of the multiplex can be in certain cases correlated. To clarify with an example, you can be a friend of the same person in two different social networks, and thus the link between you and your friend exists in both layers of an imaginary multilayer network representing your social connections. Many different measures have been proposed to quantify this topological similarity. For the unweighted case, a first, associated to a couple of layers, is called *global overlap* [49], and is defined as the total number of pairs of nodes connected at the same time by a link in both layers, or, in other words, the total number of links that are in common between layer α and layer α':

$$O^{\alpha,\alpha'} = \sum_{i<j} M_{j\alpha}^{i\alpha} M_{j\alpha'}^{i\alpha'} \qquad (3.40)$$

(for other variants on *overlapping* please refer to [34]). A second quantity, associated this time to the whole multiplex, is the *degree of multiplexity* [161], defined as the fraction of node pairs that have multiple edge types between them. This quantity is obtained by dividing the number of node pairs that have multiple edge types between them by the total number of adjacent node pairs [161]. For the specific case of weighted multilayer networks, we refer to [190] and to [53].

In particular, here we present two main weighted measures of multiplex networks – that is, *multistrength* and the *inverse multiparticipation ratio* [190]. The *multistrength* for a node i in layer α, indicated by $s_{i,\alpha}^{\bar{m}}$, is obtained by summing the weights of a certain type of multilink $\bar{m}$ – that is, the set of links connecting a given pair of nodes in the different layers of the multiplex – incident to a single node:

$$s_{i,\alpha}^{\bar{m}} = \sum_{j=1}^{N} a_{ij}^{\alpha} A_{ij}^{\bar{m}}. \qquad (3.41)$$

Given that for every layer α and node i the nontrivial multistrengths must include multilinks $\bar{M}$ with $\bar{M}_{\alpha} = 1$, the number of nontrivial multistrengths α $s_{i,\alpha}^{\bar{m}}$ is given by 2^{M-1}. It follows that the number of multistrengths that can be obtained for each node in a multiplex network of M layers is $M2^{M-1}$. The *inverse multiparticipation ratio* for a layer α $Y_{i,\alpha}^{\bar{m}}$ is used to measure the heterogeneity of the weights of multilinks incident upon a single node:

$$Y_{i,\alpha}^{\vec{m}} = \sum_{j=1}^{N} \left(\frac{a_{ij}^{\alpha} A_{ij}^{\vec{m}}}{\sum_r a_{ir}^{\alpha} A_{ir}^{\vec{m}}} \right)^2. \tag{3.42}$$

Within the context of a weighted multiplex network, the importance of considering the interacting layers in the analysis of complex system has been proven by [190], which evaluated the additional amount of information provided by weighted properties of multilinks over the one contained in single layers, further strengthening the evidence that the analysis of multilayer networks cannot be confined to the partial analysis of single layers.

3.5.3 Pairwise Multiplexity

In this last case, correlation is provided in terms of *correlated activity patterns* in the multilayer. A node i is said to be active at a layer α if $k_i^\alpha > 0$. For each node i, we can associate a node activity vector $\mathbf{b_i} = \{b_i^{[1]}, b_i^{[2]}, \ldots, b_i^{[L]}\}$ where $b_i^{[\alpha]} = 1$ if node i has at least one edge at layer α and is 0 otherwise. The node activity B_i of the node i corresponds to the total number of layers where i is active [205]:

$$B_i = \sum_{\alpha} b_i^{[\alpha]}. \tag{3.43}$$

By definition, $0 \le B_i \le L$. Analogously, the layer activity of α is given by the number of active nodes in layer α [205]:

$$N^{[\alpha]} = \sum_{i} b_i^{[\alpha]}. \tag{3.44}$$

By definition, $0 \le N^{[\alpha]} \le N$. We can now define the *layer pairwise multiplexity* $Q_{\alpha,\beta}$, which is a measure of correlation between the layers, as [205],

$$Q_{\alpha,\beta} = \frac{1}{N} \sum_{i} b_i^{[\alpha]} b_i^{[\beta]}, \tag{3.45}$$

that corresponds to the fraction of nodes that are active in both layer α and β. Examples of the distribution of the pairwise multiplexity for a continental airports network, for the papers published in the journals of the American Physical Society (APS), and for the movies in the Internet Movie Database (IMDb) are reported in Figure 30. Similarly, we can define the *node pairwise multiplexity*, measuring the correlation of activities between two nodes, as the fraction of layers in which both node i and node j are active [53, 84].

We conclude this section showing how the study of the correlation is fundamental since it alters the critical properties of the network. In fact, the pattern of correlated multiplexity (Figure 31) is crucial since it affects the multiplex

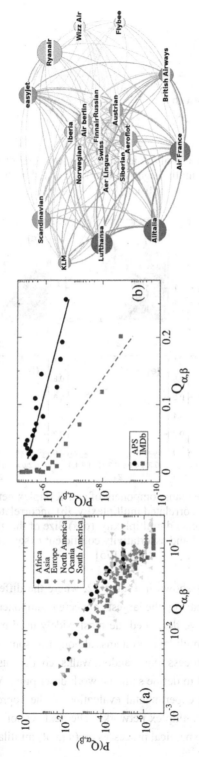

Figure 30 (a) Airline, (b) APS and IMDb networks. (c) Top 20 airlines in Europe by number of covered airports. Figure from [205].

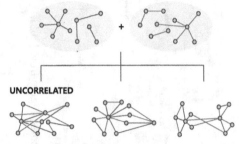

Figure 31 Different types of duplex couplings: uncorrelated, maximally positive (MP), and maximally negative (MN) correlated multiplexity. Figure from [175].

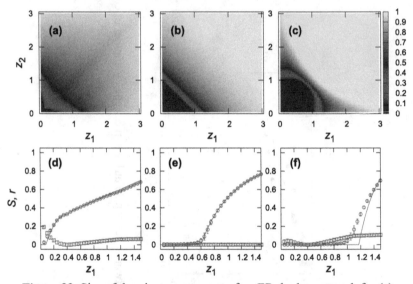

Figure 32 Size of the giant component of an ER duplex network for (a) maximally positive correlated multiplexity, (b) uncorrelated, and (c) maximally negative correlated multiplexity. (d–f) Size of the giant component (circles) for $z_2 = 0.4$ and assortativity coefficient r (squares). Figure from [175].

system's connectivity as shown in Figure 32, where the different choices of correlations have an impact on the largest connected component.

In this section, we have described the most widely used measures for the topological analysis of multilayer systems, although in some cases, we have used some dynamical process (e.g., random walks, continuous-time diffusion or shortest-path routing) to define some network descriptors. We refer to [63] for a taxonomy and an experimental evaluation of the approaches to compare different layers in multiplex networks. The next section will be entirely dedicated to introduce dynamical processes on (and of) multilayer networks.

4 Multilayer Dynamics

In the previous section, we discussed many structural descriptors of a multilayer network. However, some definitions adopted for the structural analysis are, in fact, based on some notion of dynamical process on the top of the network. In this section, the reader will have the opportunity to shed light on a spectrum of important network-driven phenomena, such as diffusive processes (Sections 4.1), synchronization dynamics (Section 4.2), cooperation dynamics (Section 4.3), intertwined/interdependent processes (Section 4.4), percolation (Section 4.5), and cascade failures (Section 4.6). Note that we refer to [206, 246] and [53, 161] for further information about the dynamics of multiplex networks, like growth processes. In the following, we will distinguish two main classes of dynamics [103] on the top of multilayer systems (see Figure 33):

- **Type–I**: *single dynamics* is defined and it is the same for all layers. Here we can find discrete [98, 167, 260, 283] and continuous-time [61, 136, 235, 258, 275] diffusion, pattern formation [24, 25, 71, 165], communicable information [116], synchronization [179, 241, 254, 255, 305], epidemic spreading of one disease [50, 69, 242, 282, 291], adoption dynamics, diffusion of innovation and other contagion processes [151, 233, 279, 292, 301], and opinion dynamics [11, 15, 21, 108, 156]. In this class we find also system

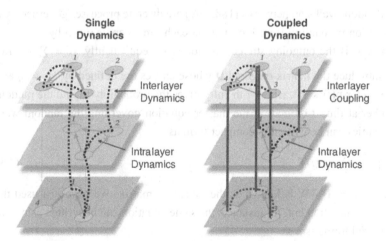

Figure 33 Two classes of dynamical processes on the top of a multilayer network. Left: single dynamics – for example, a random walk or the diffusion of some quantity – is defined for all layers. Right: coupled dynamics, where a distinct process is defined in each layer and interdependency is specified by means of coupled governing equations. Reproduced with permission from [93].

control [227, 303], congestion dynamics [77, 261, 271], cooperation processes [37, 139, 189], percolation [166, 176, 177], and cascade failures [68, 284].

- **Type–II**: *coupled dynamics*, where multiple dynamical processes are defined on each layer and their interdependency is operationally encoded into adequately coupled equations, for instance by means of interlayer connectivity when present. In this class we can find processes that combine epidemics spreading with human behavior [26, 122, 123, 141, 181, 182, 187, 286, 294, 300], simple and complex contagion [85], evolutionary game dynamics with social influence [10], cooperative with competitive epidemics spreading [81, 88, 89, 109, 249], transport with synchronization [208], interdependent human flows [264], and other interdependent processes [140].

4.1 Diffusive Processes

Diffusive processes, either discrete or continuous in time, provide reasonable models for a wide spectrum of phenomena, from random searches to population dynamics, from system exploration and navigability to information diffusion. In the following, we will consider two emblematic classes of models: random walks and continuous-time diffusion.

Random Walks on Single-Layer Networks

Random walks on networks [188, 209] are discrete processes governed by a transition probability from i to j that, at each time step, is given by $T_{ij} = \frac{W_{ij}}{s_i}$, where s_i is the outgoing strength of node i or, equivalently, $s_i = \sum_j W_{ij}$. Let us introduce the matrix diagonal $\mathbf{D}$ whose entries are defined by $D_{ii} = s_i$ and the vector $\mathbf{p}(t) \in \mathbb{R}^N$, whose i-th entry gives the probability to find the random walker at time t on node i: the master equation governing the random walk dynamics can be written in compact form as

$$\mathbf{p}(t+1) = \mathbf{p}(t)\mathbf{T}, \tag{4.1}$$

where $\mathbf{T} = \mathbf{D}^{-1}\mathbf{W}$ is known as the transition matrix. We have just used the standard notation but, for instance, the same equation can be written using our tensorial formalism as

$$p_j(t+1) = p_i(t)T^i_j, \tag{4.2}$$

where Einstein convention (sum over i) is used and i,j indicate covariant and contravariant indices, not the entries of the corresponding rank-1 and rank-2 tensors.

The reader familiar with the Markov chains formalism will immediately recognize the continuous-time Markov chain on N states (i.e., on the set of nodes), with independent exponential holding times with parameter $\lambda = 1$ and jumping probabilities given by the transition probabilities of the discrete-time random walk, and that can be equivalently defined by the forward equation

$$\dot{\mathbf{p}}(t) = -\mathbf{p}(t)\tilde{\mathbf{L}}, \tag{4.3}$$

where $\tilde{\mathbf{L}} = \mathbf{I} - \mathbf{D}^{-1}\mathbf{W}$ is known as the random walk normalized Laplacian and $-\tilde{\mathbf{L}}$ is the generator of the continuous-time process with an initial condition $\mathbf{p}(0) = \mathbf{p}_0$. Equivalently, in tensorial notation we can rewrite the equation and initial condition as

$$\begin{cases} \dot{p}_j(t) = -\sum_{i=1}^{N} \tilde{L}_j^i p_i(t) \\ p_i(0) = q_i \end{cases}, \tag{4.4}$$

with $\tilde{L}_j^i = \delta_j^i - T_j^i$ indicating the component (i,j) of the random walk normalized Laplacian tensor, δ_j^i the Kronecker delta, and T_j^i the component (i,j) of the transition probability tensor. The solution of this equation is given by

$$\mathbf{p}(t) = \mathbf{p}_0 e^{-t\tilde{\mathbf{L}}}, \tag{4.5}$$

where the intervening exponential matrix is known as the *propagator of the dynamics*. Note that the transition matrix can encode other rules governing the jumps of a random walker from one node to another: by changing the *flavor* of the walk, we can explore a broad spectrum of stochastic processes, as we will see in the next sections, where we will generalize these dynamics to multilayer networks, where a walker can jump between nodes but can also switch between layers. In general, the transition matrix is replaced by a rank-4 transition tensor $T_{j\beta}^{i\alpha}$, which governs the probability that a walker on node i in layer α will move to node j in layer β. An illustration of a random walk on the top of such a system is shown in Figure 34, while a schematic representation of the supra-adjacency matrices corresponding to the multilayer systems considered in the reminder of this section is shown in Figure 9, and is useful to get some visual insights about the structure of the corresponding supra-transition matrices.

Random Walks on Edge-Colored Multigraphs

On this class of systems (note that $M_{j\beta}^{i\alpha} = 0$ for all $\alpha \neq \beta$), it is possible to define random walk dynamics in two distinct ways. On one hand, we can allow the walker to jump through edges with different colors, which are considered as multiple edges, regardless of their colors, where a node's degree is the sum of

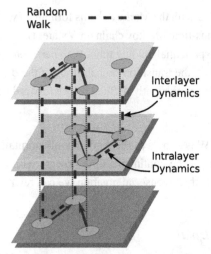

Figure 34 A random walker visits the nodes of a multilayer network: starting from one state node in a specific layer, it is allowed to (i) stay in the same state node, (ii) jump to a neighbor within the same layer, (iii) switch to another state node on another layer, or (iv) jump elsewhere while subjected to the transition tensor $T^{i\alpha}_{j\beta}$. The walk is defined by the ordered sequence of visited intra- and interlayer connections. Reproduced with permission from [93].

its state nodes' degree [35]. Operationally, this approach is equivalent to performing a classical random walk on the top of the aggregate representation of the system: in general, this could not be desirable because it is difficult to quantify the effects of neglected information (i.e., connectivity patterns per layer) on the dynamics.

On the other hand, we can allow walkers to explore each layers independently and then integrate the transition matrices governing their dynamics [131] as

$$\langle T^i_j \rangle = \sum_{\alpha=1}^{L} m_{i,\alpha} T^{i\alpha}_{j\alpha}, \qquad \sum_{\alpha=1}^{L} m_{i,\alpha} = 1 \quad \forall i \in V, \tag{4.6}$$

where $m_{i,\alpha} \geq 0$ are weights enabling us to tune the importance of each layer, relative to node i in the final dynamics of the system. If $m_{i,\alpha} = \frac{1}{L}$ for all i and α, then $\langle T^i_j \rangle$ is simply the average transition matrix over all layers. However, if we choose $m_{i,\alpha} = \frac{1}{\mu_i} 1_{\{s_i(\alpha) \neq 0\}}$, with $s_i(\alpha)$ being the out-strength of vertex i in layer α and $\mu_i = \sum_{\alpha=1}^{L} 1_{\{s_i(\alpha) \neq 0\}}$ the multiplicity of node i – that is, the number of layers in which i is not isolated – then we can discard the effect of i being isolated in one or more layers. This approach preserves the diversity of information coming from layers' connectivity patterns and allows for more flexibility. In fact, in Equation (4.6), we use a finite mixture of probability mass functions with weights $m_{i,\alpha} \geq 0$ such that $\sum\limits_{\alpha=1}^{L} m_{i,\alpha} = 1$, but if additional information from the data is available, then the transition matrices in Equation (4.6) can be combined by means of weights encoding the relative importance given to each layer, thus providing an operational way to prioritize the available information from specific layers. Here, for the sake of simplicity, we are just considering that all layers are equally important and adopting the weights $m_{i,\alpha} = \frac{1}{\mu_i} 1_{\{s_i(\alpha)\neq 0\}}$.

Random Walks on Multilayer Networks

In the following, we indicate by $D(i; \alpha, \beta)$ the interlayer connections between replicas across nodes – that is, the entries of $M_{j\beta}^{i\alpha}$ for $i = j$. In this framework, the outgoing strength of node i in layer α is given by $s_i(\alpha) = \sum\limits_{j=1}^{N} M_{j\alpha}^{i\alpha}$, whereas the multilayer strength – discarding the interlayer edges – is given by $s_i = \sum\limits_{\alpha} s_i(\alpha) = \sum\limits_{\alpha=1}^{L} \sum\limits_{j=1}^{N} M_{j\alpha}^{i\alpha}$. The interlayer strengths are similarly given by $S_i(\alpha) = \sum\limits_{j} \sum\limits_{\beta\neq\alpha} M_{j\beta}^{i\alpha}$ and, consequently, $(s_i(\alpha) + S_i(\alpha))_{i=1,\alpha=1}^{N,L}$ is the out-strength supra-vector with NL components obtained as the row sums of the supra-adjacency matrix. The reader will notice that the same results can be simply obtained by using the tensorial formalism and Einstein convention as $s^{i\alpha} = M_{j\beta}^{i\alpha} u^j u^\beta$.

Usually, for the sake of simplicity, isolated nodes and components, as well as more complex patterns of interlayer connectivity, are discarded from the analysis – for example, focusing on the largest connected component and considering only nodes that exist in all layers, since a more general framework has a non-negligible impact on the calculation of transition probabilities. In general, we can overcome these limitations, typical of real-world systems, by adding state nodes as isolated and adequately accounting for them in calculations. In the following, we consider three cases:

1. $S_i(\alpha) = 0$ – that is, interlayer connectivity is missing for node i;
2. $s_i(\alpha) = 0$ – that is, intralayer connectivity is missing for node i in layer α (i.e., the node is isolated); and
3. $s_i = 0$ – that is, the node is isolated everywhere.

In the last case, we can simply exclude the node from the analysis, since it is not part of the network in practice. Therefore we can safely assume that $s_i > 0$ for all nodes. Concerning cases 1 and 2, they require special attention only if $S_i(\alpha)+s_i(\alpha) = 0$ for some α, since the state node (i, α) is an absorbing state – that is, the walker reaching that node will stay there with probability equal to 1. To avoid absorbing states of this type, one option is to add a *teleportation* dynamics

– that is, a uniform but small probability equal to $\frac{1}{NL}$ – to jump elsewhere in the system even if outgoing connections are missing. This approach has been successfully used in Google's PageRank to rank web pages [62], as well as to identify functional modules in monoplex [238] and multilayer networks [99]. Here, we use a similar approach, with the effect that the occupation probability of the state node (i, α) is effectively reduced. Once the transition tensor is well defined, as we will see later in this section for some specific processes, we can define the master equation in discrete time as

$$p_{j\beta}(t+1) = \underbrace{T_{j\beta}^{i\beta} p_{j\beta}(t)}_{\text{stay}} + \underbrace{\sum_{\substack{\alpha=1 \\ \alpha \neq \beta}}^{L} T_{j\beta}^{j\alpha} p_{j\alpha}(t)}_{\text{switch}} + \underbrace{\sum_{\substack{i=1 \\ i \neq j}}^{N} T_{j\beta}^{i\beta} p_{i\beta}(t)}_{\text{jump}} + \underbrace{\sum_{\substack{\alpha=1 \\ \alpha \neq \beta}}^{L} \sum_{\substack{i=1 \\ i \neq j}}^{N} T_{j\beta}^{i\alpha} p_{i\alpha}(t)}_{\text{switch and jump}},$$

where $p_{j\beta}(t)$ indicates the probability of finding a random walker in node j of layer β at time t, and the contributions of jumps and switches are made explicit. Note that using the tensorial formalism, this expression would be compressed into the elegant master equation

$$p_{j\beta}(t+1) = p^{i\alpha}(t) T_{i\alpha}^{j\beta}, \tag{4.7}$$

with the continuous-time version described by the forward equation

$$\dot{p}_{j\beta}(t) = -\tilde{L}_{j\beta}^{i\alpha} p_{i\alpha}(t), \tag{4.8}$$

where $\tilde{L}_{j\beta}^{i\alpha} = \delta_{j\beta}^{i\alpha} - T_{j\beta}^{i\alpha}$ is the random walk normalized Laplacian tensor.

It is important to remark that a random walk dynamics is affected by both structure and transition mechanisms – that is, by fixing the topology we can always define distinct transition rules encoding different stochastic movements between nodes and exploration strategies, which we have previously named flavors.

As mentioned before, in PageRank random walks (PRRWs), a teleportation (or nonlocal jumping) parameter tunes the ability of the random walker to escape from absorbing states and to reach nodes that are not in the neighborhood of the current node [62, 102]. The random walk introduced in the previous section for monoplex networks is known as a classical random walk (CRW) [97, 98]. There are several walks that can be defined on multilayer networks, which generalize their monoplex counterparts or are specific to multilayer systems. For instance, we can define multilayer diffusive random walks (DRW), generalizing the one defined in [98]; maximal-entropy random walks (MERW) [98], generalizing monoplex walks that localize around topological defects [70]; and a physical random walk with relaxation (PrRW) [99]. Transition probabilities for these processes are reported in Table 1.

Table 1 Entries of the transition tensor for distinct random walks mentioned in the text, namely classical (CRW), PageRank (PRRW), diffusive (DRW), maximal-entropy (MERW), and physical with relaxation (PrRW) random walks. To keep the notation simple, we use $\sigma_j(\beta) = s_j(\beta) + S_j(\beta)$ and $s_{max} = \max_{i,\alpha}\{\sigma_i(\alpha)\}$ for the outgoing strengths (see the text for details). Note that the teleportation parameter is usually denoted by α: to avoid confusion; here we indicate it by r. Finally, λ_{max} is the largest eigenvalue of the multilayer adjacency tensor (see Section 3.3.3). It is worth remarking that for CRW, DRW, and MERW, these transition rules generalize to any multilayer system such as the ones introduced in [98] for multiplex networks; PrRW is defined as in [99]; PRRW generalizes the walk introduced in [102]. Reprinted table with permission from [46]. Copyright (2021) by the American Physical Society.

	CRW	PRRW	DRW	MERW	PrRW
$T^{i\beta}_{j\beta}$	$\dfrac{M^{i\beta}_{j\beta}}{\sigma_j(\beta)}$	$r\dfrac{M^{i\beta}_{j\beta}}{\sigma_j(\beta)}+\dfrac{1-r}{NL}$	$\dfrac{s_{max}+M^{i\beta}_{j\beta}-\sigma_j(\beta)}{s_{max}}$	$\dfrac{M^{i\beta}_{j\beta}}{\lambda_{max}}$	$(1-r)\dfrac{M^{i\beta}_{j\beta}}{s_j(\beta)}+r\dfrac{M^{i\beta}_{j\beta}}{s_j}$
$T^{i\alpha}_{j\beta}$	$\dfrac{M^{i\alpha}_{j\beta}}{\sigma_j(\alpha)}$	$r\dfrac{M^{i\alpha}_{j\beta}}{\sigma_j(\alpha)}+\dfrac{1-r}{NL}$	$\dfrac{M^{i\alpha}_{j\beta}}{s_{max}}$	$\dfrac{M^{i\alpha}_{j\beta}}{\lambda_{max}}\dfrac{V_{j\beta}}{V_{j\alpha}}$	$r\dfrac{M^{i\beta}_{j\beta}}{s_j}$
$T^{i\beta}_{j\beta}$	$\dfrac{M^{i\beta}_{j\beta}}{\sigma_i(\beta)}$	$r\dfrac{M^{i\beta}_{j\beta}}{\sigma_i(\beta)}+\dfrac{1-r}{NL}$	$\dfrac{M^{i\beta}_{j\beta}}{s_{max}}$	$\dfrac{M^{i\beta}_{j\beta}}{\lambda_{max}}\dfrac{V_{j\beta}}{V_{i\beta}}$	$(1-r)\dfrac{M^{i\beta}_{j\beta}}{s_i(\beta)}+r\dfrac{M^{i\beta}_{j\beta}}{s_i}$
$T^{i\alpha}_{j\beta}$	$\dfrac{M^{i\alpha}_{j\beta}}{\sigma_i(\alpha)}$	$r\dfrac{M^{i\alpha}_{j\beta}}{\sigma_i(\alpha)}+\dfrac{1-r}{NL}$	$\dfrac{M^{i\beta}_{j\beta}}{s_{max}}$	$\dfrac{M^{i\alpha}_{j\beta}}{\lambda_{max}}\dfrac{V_{j\beta}}{V_{i\alpha}}$	$r\dfrac{M^{i\beta}_{j\beta}}{s_i}$

We conclude this section with a short discussion about the physical random walk, introduced in [98] to describe those dynamics where the state nodes have a "common memory" such that information reaching one state node is *instantaneously* diffused to all state nodes of the corresponding physical node. This mechanism encodes a variety of processes in the real world. For instance, in a network of digital interactions among individuals, such as social media platforms, we can get a rumor in a particular platform through our intralayer connections, but at the same time that rumor will be known to our alter egos in other social media platforms. In this scenario, interlayer connectivity does not carry physical meaning and consequently is ignored. The physical random walk with relaxation (PrRW) [99] can be seen as a variant of PRW, where the knowledge of interlayer connectivity is not required and is therefore discarded: from Table 1 it is easy to identify that its transition probabilities balance intra- and interlinks by means of probability weights $1 - r$ and r, respectively.

It is worth remarking here that several other types of walks can be defined, including ones describing quantum processes. See [48] for a review of this topic. From a mathematical perspective, we can write a master equation similar in shape to the one of a CRW:

$$|\dot{\psi}\rangle = -i\mathbf{L}_Q|\psi\rangle, \tag{4.9}$$

where i is the imaginary unit and $\mathbf{L}_Q = \mathbf{D}^{-1/2}\mathbf{L}\mathbf{D}^{-1/2}$, $\mathbf{L}$ being the combinatorial Laplacian matrix. The evolution equation for this type of continuous-time quantum walk is time-reversible and, unlike CRWs, quantum ones: (i) do not admit a stationary distribution and (ii) are deterministic in their dynamic nature and stochastic with respect to measurements [60]. In the future, it will be of interest to study the emergence of potentially unseen physical properties related to quantum walks on the top of multilayer networks.

Continuous-Time Diffusion

Let us assume we have some quantity – for example, water – free to flow in a network of nodes through pipes, which can be represented by links between pairs of nodes. Given an initial distribution, what is the level of this quantity in each node at a given time t? This problem can be cast into continuous-time diffusion on a network.

In the following, let us indicate by $x_i(t)$ the state vector, in tensorial formalism, carrying information about the quantity in each node of a monoplex network at time t. Let $x_i(0)$ be the initial state vector: its evolution over time can be modeled by the diffusion equation

$$\dot{x}_j(t) = \mathcal{D}\left[W_j^i x_i(t) - W_k^i u_i e^k(j) x_j(t)\right], \tag{4.10}$$

where $\mathcal{D}$ is a diffusion constant, u_i is the vector of ones, and $e^k(j)$ is a canonical vector (see Section 2.1). Since $s_k = W_k^i u_i$ is the outgoing strength and $s_k e^k(j) x_j(t) = s_k e^k(j)\delta_j^i x_i(t)$, the diffusion equation can be written in more compact form as:

$$\frac{dx_j(t)}{dt} = -\mathcal{D}L_j^i x_i(t), \tag{4.11}$$

where $L_j^i = W_k^l u_l e^k(j)\delta_j^i - W_j^i$ is the combinatorial Laplacian tensor [78]. It is worth remarking here that this tensor differs from the normalized random walk Laplacian introduced in the previous section, although in some cases – as for a CRW – they are related by the simple relationship $\tilde{L}_j^i = (D^{-1})_k^i L_j^k$. The solution of Equation (4.11) is given by $x_j(t) = x_i(0)e^{-\mathcal{D}L_j^i t}$, similarly to what we have seen for continuous-time random walks. However, at variance with random walks, it can be shown that in the stationary regime, the solution takes the form $x_j(\infty) \propto u_j$ – that is, one will find exactly the same fraction of the quantity in each node, uniformly distributed.

We might wonder how fast diffusion happens. Since the Laplacian matrix is semi-positive definite, it is possible to show that it can be decomposed as

$$L_j^i = Q_h^i \Lambda_k^h (Q^{-1})_j^k, \tag{4.12}$$

where Q_h^i is a matrix whose columns are the eigenvectors of the Laplacian matrix and Λ_k^h is a diagonal matrix whose entries are the Laplacian's eigenvalues. This algebraic feature allows us to prove that $e^{-\mathcal{D}L_j^i t} = Q_h^i (e^{-\mathcal{D}\Lambda t})_k^h (Q^{-1})_j^k$, highlighting that the exponential decay of $x_j(t)$ is dominated by the smallest positive eigenvalue, which usually is indicated by Λ_2. Therefore, the diffusion temporal scale is given by $\tau \approx 1/\Lambda_2$.

In the case of interconnected multilayer networks, the diffusion equation has been generalized by means of the supra-adjacency matrix [136] and the tensorial formulation [97]. In this new setup, a quantity can diffuse through interlayer connections as well. If we indicate by $X_{i\alpha}(t)$ the rank-2 state tensor at time t, then the multilayer diffusion equation can be written as

$$\frac{dX_{j\beta}(t)}{dt} = M_{j\beta}^{i\alpha} X_{i\alpha}(t) - M_{k\gamma}^{i\alpha} U_{i\alpha} E^{k\gamma}(i\beta) X_{i\beta}(t), \tag{4.13}$$

where $U_{i\alpha} = u_i u_\alpha$ and $E^{k\gamma}(i\beta) = e^k(i)e^\gamma(\beta)$. If we define the multilayer combinatorial Laplacian tensor as

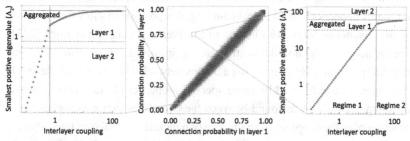

Figure 35 Continuous-time diffusion on a duplex network – that is, a multiplex consisting of two layers, obtained from two Erdős–Rényi networks with independently wiring probabilities $p_1, p_2 \in [0, 1]$, used to connect two pairs of nodes within the same layer. Diffusion speed is quantified by the second smallest eigenvalue of the Laplacian tensor, Λ_2: depending on the wiring probabilities, diffusion in the duplex can be faster (left-hand side panel) or slower (right-hand side panel) than in each layer separately. The condition for enhanced diffusion to happen is given by $\Lambda_2^{\text{multiplex}} \geq \max\{\Lambda_2^{\text{layer 1}}, \Lambda_2^{\text{layer 2}}\}$: a sharp change in the behavior of the characteristic temporal scale can be observed for varying weight of the interlayer connections (left and right panels), with a clear transition between two distinct regimes above a certain critical value of interlayer coupling. The middle panel shows when the enhanced diffusion condition holds (encoded by colors), while varying the wiring probabilities. Figure from [103].

$$L_{j\beta}^{i\alpha} = M_{k\gamma}^{l\epsilon} U_{l\epsilon} E^{k\gamma}(j\beta)\delta_{j\beta}^{i\alpha} - M_{j\beta}^{i\alpha}, \tag{4.14}$$

the diffusion equation can be written more compactly as

$$\frac{dX_{j\beta}(t)}{dt} = -L_{j\beta}^{i\alpha}X_{i\alpha}(t), \tag{4.15}$$

whose solution is given by $X_{j\beta}(t) = X_{i\alpha}(0)e^{-L_{j\beta}^{i\alpha}t}$, a clear generalization of the result obtained in the case of single-layer networks. Also in this case, the second smallest eigenvalue Λ_2 – calculated from the supra-adjacency matrix representation – governs the speed of diffusion [97, 136, 258], leading to interesting phenomena (see Figure 35 for details).

Topological Transition with Diffusive Processes

An important question concerns the emergence of such effects in spreading processes, but not only. One can adapt the dynamical rules of models defined in monolayers to encompass the fact that the topology is more complex and this, as we will see, can lead to substantially different behaviors than those observed in isolated networks. Another point of view, though, is to ask why and when a certain model could effectively behave as in an isolated network even though it runs in a layered structure, or in the other way around, why and when the multilayer dimension dominates, thus disregarding any role of the individual networks of the layers. In some special cases, such as interconnected multiplex networks with identical coupling, it is possible to show the existence of two distinct regimes as a function of the interlayer coupling strength [231], highlighting how the multilayer structure can influence the outcome of several physical processes. By considering a duplex network – that is, a multiplex with two layers, A and B – and by using spectral properties of multilayer systems, an abrupt transition was found between the two aforementioned regimes as a function of p, the coupling strength. The two regimes are inferred by analyzing the behavior of eigenvectors and eigenvalues of the supra-Laplacian matrix and are clearly distinguishable, separated by a critical point p^*. For $p \leq p^*$, the second smallest eigenvalue[11] λ_2, which is associated with a plethora of network properties, is independent of the structure of the layers and hence the dynamical processes can be studied separately, while in the regime $p > p^*$, λ_2 tends to a value independent of p – that is, depending only on the details of the intralayer networks.

The eigenvector $|v\rangle$ associated with $\lambda_2(p)$ can be split into $|v_A\rangle$ and $|v_B\rangle$, which correspond to the elements of $|v\rangle$ associated with nodes of networks A

---

[11] Note that here we are using λ_2 instead of Λ_2, as in the previous section, to keep the same notation of the original paper by [231].

and B, respectively. It can be proved [231] that for $p \leq p^*$, $|v_A\rangle = -|v_B\rangle = 0$ holds, where $|v_A\rangle = \pm\frac{1}{\sqrt{2N}}|1\rangle$, while for $p > p^*$ we have $\langle v_A|1\rangle = \langle v_B|1\rangle = 0$. Physically, this means that in the subcritical regime, the layers are structurally independent whereas in the supercritical regime, the interlayer connection dominates, imposing the same sign in the eigenvector for nodes across networks and alternating the sign for nodes in the same layer. Therefore, the algebraic phase transition can be visualized in several ways. Panels in Figure 36 show the behavior of the eigenvectors in the subcritical and supercritical regions. Here, $\lambda_2(p)$ displays a singular point p^* at which the first derivative is not continuous, a sign of an abrupt transition. For $p \leq p^*$, λ_2 grows as $2p$, while in the other regime, it tends to the value that would take for a weighted superposition of the two layers A and B, whose Laplacian is $(1/2)(\mathcal{L}_A + \mathcal{L}_B)$. This abruptness can be observed more directly in the middle and lower panels, which show the behavior of $\langle v_A|v_B\rangle$, $\langle v_A|1\rangle$, and $\langle v_B|1\rangle$ as a function of p.

4.2 Synchronization Processes

Synchronization is an emergent phenomenon of a population of dynamically interacting units that, usually with a second-order phase transition [110, 199, 267], start operating in a collective, coherent way. Synchronization phenomena may be found in biology, sociology, and ecology and include birds flocking, fireflies flashing, people singing, and neurons spiking, just to mention a few examples. Once a fully synchronized state is reached, the (linear) stability of such a state is tested by studying the effects of a small perturbation of the system state. This approach was introduced in [219] for simple network configurations and has been widely used and extended to complex network topologies. Here we briefly introduce the framework of the master stability function (MSF) for a network of oscillators and then extend the formalism to multilayer networks. For a more exhaustive description, see [17, 54].

It is worth remarking that, in the following, we will use the more traditional vector notation where $\mathbf{x}(t)$ indicates the state of the system at time t. In some cases, we will also use the Kronecker product operator $\otimes$, as is usual in equations governing synchronization dynamics. Our choice is to help the reader link these concepts to the original studies that introduced them. Nevertheless, the whole section could consider the tensorial formalism, where the system state is indicated by the rank-1 tensor $x_\ell(t)$ and where products such as $\mathbf{A} \otimes \mathbf{B}$ are indicated as $A_\beta^\alpha B_\delta^\gamma$.

Let us consider a network of N identical oscillators in an m-dimensional space, where, in the absence of any interaction, the dynamics of each node i is described by:

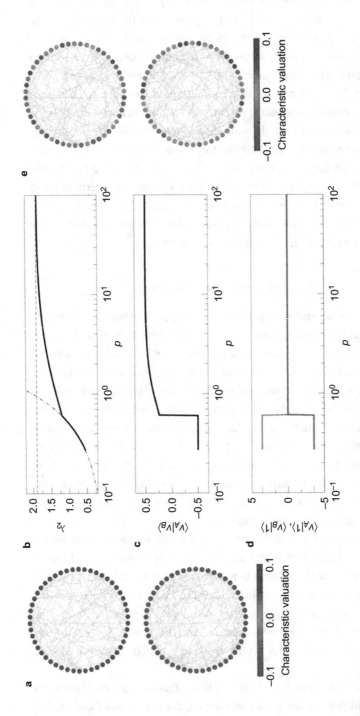

Figure 36 Algebraic phase transition. In (a) and (e), the intralayer connectivity of a duplex (i.e., a multiplex consisting of two layers) is shown, where the color of each node refers to the value of the corresponding component in the Fiedler vector – that is, the eigenvector associated with the second smallest eigenvalue λ_2. In (a) $p = p^* = 0.602$ and in (e) $p = 0.603$. In the middle column is displayed the eigenvalue λ_2 (b), the inner product $\langle v_A | v_B \rangle$ (c), and, in (d), inner products of $|v_A\rangle$ and $|v_B\rangle$ with the unit vector – that is, the sum of their elements (see the text for details). Figure reproduced from [231].

$$\dot{\mathbf{x}}_i = \mathbf{F}(\mathbf{x}_i), \qquad i = 1, 2, \ldots, N; \, \mathbf{x}_i \in \mathbb{R}^m. \tag{4.16}$$

We introduce an interaction between oscillators due to the fact that they are coupled in an unweighted network specified by the adjacency matrix $\mathbf{A} = \{A_{ij}\}$ and we define the output function $\mathbf{H}(\mathbf{x})$ as the function that governs the interaction between nodes. We also assume that the coupling between the oscillators is diffusive – that is, the effect that node j has on node i is proportional to the difference between $\mathbf{H}(\mathbf{x}_j)$ and $\mathbf{H}(\mathbf{x}_i)$. Then, the evolution of the state of node i is given by:

$$\dot{\mathbf{x}}_i = \mathbf{F}(\mathbf{x}_i) + \sigma \sum_{i=1}^{N} A_{ij}[\mathbf{H}(\mathbf{x}_j) - \mathbf{H}(\mathbf{x}_i)] = \mathbf{F}(\mathbf{x}_i) - \sigma \sum_{j=1}^{N} L_{ij}\mathbf{H}(\mathbf{x}_j), \tag{4.17}$$

where $\mathbf{L}$ is the Laplacian matrix and σ is the coupling strength.

Stability of Synchronized States

The MSF approach to test the stability of a fully synchronized state, described in Appendix A, was extended to multilayer complex systems [105]. In this work, it is considered a network with M different layers, each layer representing a different kind of interaction between nodes. Equation (4.17) is thus extended to describe the dynamics of the whole system:

$$\dot{\mathbf{x}}_i = \mathbf{F}(\mathbf{x}_i) - \sum_{\alpha=1}^{M} \sigma_\alpha \sum_{j=1}^{N} L_{ij}^{(\alpha)}\mathbf{H}^{(\alpha)}(\mathbf{x}_j), \tag{4.18}$$

where α is the index accounting for layers. We obtain the M-parameter equation describing the time evolution of perturbation error:

$$\dot{\xi}_i = \left[J\mathbf{F}(\mathbf{s}) - \sum_{\alpha=1}^{M} \sigma_\alpha \lambda_i^{(\alpha)} J\mathbf{H}_\alpha(\mathbf{s}) \right] \xi_i, \tag{4.19}$$

where ξ_i is the eigenmode associated with the eigenvalue λ_i of $\mathbf{L}$. As in the case of a dynamics evolving on top of a single layer, in a multilayer system, the stability of the synchronized state is completely specified by the sign of the maximum conditional Lyapunov exponent Λ_{max}. In particular, it is also found that stability of the complete synchronization state may be reached even if each layer, taken individually, is unstable: a very interesting feature for practical application.

It is worth noting that, together with complete synchronization, a network may exhibit other forms of synchronization where clusters of nodes have a synchronized dynamics but different clusters evolve on distinct time evolutions. This type of synchronization is called cluster synchronization (CS) and it has been well studied in terms of cluster formation, stability, and role of network symmetries [207, 220, 265]. Clustered synchronization has been also studied in multiplex [155] and, more recently, in multilayer [107] networks, and cluster stability has been tested as a function of intra- and interlayer symmetries. To describe the evolution of the perturbation error for complex synchronization patterns, such as in CS, it is useful to rewrite Equation (A.1) with a more compact formalism. To this end, we write the state vector as a vector of vectors, $\mathbf{X} = (\mathbf{x}_1; \mathbf{x}_2; \ldots; \mathbf{x}_N)$, and the variational equation assumes the form:

$$\delta\dot{\mathbf{X}} = [\mathbf{I}_N \otimes J\mathbf{F}(\mathbf{s}) - \sigma\mathbf{L} \otimes J\mathbf{H}(\mathbf{s})] \, \delta\mathbf{X}, \tag{4.20}$$

where $\mathbf{I}_N$ is the identity matrix and $\otimes$ is the Kronecker product. Equation (4.20) can be decoupled into N independent equations by diagonalizing $\mathbf{L}$. However, to deal with CS and multilayer interactions, we have to further generalize Equation (4.20). The variational equations for complex synchronization patterns on generalized networks has the form [306]:

$$\delta\dot{\mathbf{X}} = \left[\sum_{l=1}^{L} \mathbf{D}^{(l)} \otimes J\mathbf{F}(\mathbf{s}^l) - \sum_{l=1}^{L} \sum_{\alpha=1}^{M} \sigma_\alpha \mathbf{L}^{(\alpha)} \mathbf{D}^{(l)} \otimes J\mathbf{H}^{(\alpha)}(\mathbf{s}^l) \right] \delta\mathbf{X}, \tag{4.21}$$

where the identity matrix has been replaced by the diagonal matrix $\mathbf{D}^{(l)}$, whose generic element $D_{ii}^{(l)} = 1$ if node i belongs to lth dynamical cluster and $D_{ii}^{(l)} = 0$ otherwise.

In a recent paper [306], it was established that to optimally decouple Equation (4.21), the matrices encoding the synchronization pattern and the interaction pattern – that is, $\{\mathbf{D}^{(l)}\}$ and $\{\mathbf{L}^{(\alpha)}\}$ – should be simultaneously diagonalized. In this work, an algorithm was also developed to find the finest simultaneous block diagonalization.

Synchronization in a Network of Phase Oscillators

One of the first approaches to describe phase synchronization in an ensemble of oscillators on multiplex networks was done in 2015 [128], to investigate the synchronization of indirectly coupled units using a system composed by two layers, where the top layer was made of disconnected oscillators and the bottom one, modeling the medium, consisted of oscillators coupled according to a given topology and with a characteristic natural frequency. Each node of the multiplex was modeled as a Stuart-Landau (SL) oscillator – that is, an oscillator with amplitude as well as phase dynamics. Therefore, the Kuramoto model (KM) (see Appendix B) can be retrieved as a limiting case when the amplitude dynamics vanishes. The Kuramoto order parameter can be generalized to the multiplex framework as:

$$r_{ij}^{\alpha\beta} = \left| \left\langle e^{i[\theta_i^\alpha(t) - \theta_j^\beta(t)]} \right\rangle_t \right|, \tag{4.22}$$

while intra- and interlayer coherence, respectively, can be also defined by

$$r^\alpha = \frac{1}{N(N-1)} \sum_{i,j=1}^{N} r_{ij}^{\alpha\alpha},$$

(4.23)

and

$$r^{\alpha\beta} = \frac{1}{N} \sum_{j=1}^{N} r_{jj}^{\alpha\beta}.$$

(4.24)

By studying a population of N disconnected oscillators, indirectly coupled through an inhomogeneous medium, authors have shown the onset of intralayer synchronization without interlayer coherence – that is, a state in which the nodes of a layer are synchronized between them without being synchronized with those of the other layer (see Figure 37, left panel). Synchronization of units that are not connected requires the presence of an amplitude dynamics as the regime of intralayer synchronization is not observed in purely phase oscillators, such as those in the KM.

As previously mentioned, not only were continuous second-order transitions observed in an ensemble of networked phase oscillators, but examples of an abrupt first-order transition, named explosive synchronization (ES), were also observed [137, 178]. It was first also observed in a network of oscillators presenting a positive correlation between natural frequencies and the degree of the nodes. Moreover, ES was also studied in systems where a *local-order* parameter for the i-th oscillator is defined [305]:

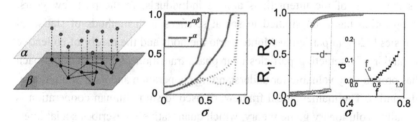

Figure 37 Left: *Two-layer* multiplex network with one-to-one coupling between the layers. In the top layer (α), the nodes only interact with those in the bottom one, whereas in the bottom layer (β), the nodes also interact with other members of the same layer. Middle: Kuramoto-order parameters (r^α) and ($r^{\alpha\beta}$) versus coupling coefficients $\lambda = \lambda_{\alpha\beta} = \lambda_\beta/5$. Continuous lines refer to a multilayer network of Stuart-Landau oscillators with $a = 1$, whereas the dashed ones refer purely phase oscillators ($a \to \infty$). Right: Synchronization transitions in two-layer networks with a fraction of the nodes adaptively controlled by a local-order parameter $f = 1$. Squares and circles (triangles and stars) refer to the values of $R1$ ($R2$), and the insets show the corresponding dependence of the width of a hysteretic loop d on f. Left and middle panels readapted from [128], right panel from [305].

$$r_i(t)e^{i\Phi(t)} = \frac{1}{k_i} \sum_{j=1}^{k_i} \sin(\theta_j), \tag{4.25}$$

and where the phase dynamics is expressed as:

$$\dot{\theta}_i = \omega_i + \sigma\alpha_i \sum_{j=1}^{N} A_{ij} \sin(\theta_j - \theta_i). \tag{4.26}$$

The overall amount of phase coherence in the network is measured by means of the global-order parameter R:

$$Re^{i\Psi} = \frac{1}{N} \sum_{j=1}^{N} e^{i\theta_j}, \tag{4.27}$$

where $0 \leq R \leq 1$ and Ψ denotes the average phase.

Explosive synchronization onset has been reported in a system of two interdependent networks (see Figure 37, right panel), with the same size and where nodes on the two layers are coupled in a one-to-one correspondence, so that a group of oscillators in the first layer is controlled by the local-order parameters of the corresponding nodes on the second layer, and vice versa [305]. Nevertheless, it was shown that ES is a property of a generic multilayer network as long as some microscopic suppressive rule can prevent the formation of the giant synchronization cluster that characterizes second-order transitions.

4.3 Game Dynamics: Cooperation Processes

Human cooperation may be intended as a collective behavior that emerges as the result of the interactions among individuals. In the past few years, cooperation has been studied in social sciences with methods of statistical physics [224], in particular Monte Carlo methods and the theory of collective behavior of interacting particles near phase-transition points. That approach has proven very valuable for understanding cooperation and its spatiotemporal dynamics. The mathematical framework used to study human cooperation is usually evolutionary game theory, which quantitatively describes social interactions using example games and formalizes the concept of the social dilemma, intended as the conflictual choice that an individual has between doing what is best for society or doing what is best for themselves.

Evolution of cooperation strongly depends on the population interaction network: individuals who have the same strategy are more likely to interact [211, 212], effectively creating resilient cooperative clusters in a structured population, a phenomenon named network reciprocity. When the edges that determine the interaction among individuals were fixed on a time scale, it was demonstrated [247] that in heterogeneous populations – modeled by networks with degree distributions exhibiting a power-law behavior – the sustainability of cooperation is simpler to achieve than in homogeneously structured populations. Here we address the problem of how human cooperation emerges in multiplex networks, with different interaction layers that can account for different kinds of social ties an individual may be involved in.

In particular, we consider a prisoner's dilemma game implemented in a set of M interdependent networks, characterized by an average degree $\langle k \rangle$, each of them containing the same number N of nodes. Each individual is represented by one node in each of the layers and we define a set of adjacency matrices $\{A^\alpha\}$ so that $A_{ij}^\alpha = 1$ when two nodes are connected in layer α and $A_{ij}^\alpha = 0$ otherwise.

At each time step, for each of the k_i^α games played, each individual i facing a cooperator collects a payoff $\pi_{ij} = R$ or $\pi_{ij} = T$ when playing as a cooperator or as a defector, respectively. Conversely, if i faces a defector, i collects a payoff $\pi_{ij} = S$ or $\pi_{ij} = P$ playing as a cooperator or as a defector, respectively. For game parameters, it holds that $S < P < R < T$. The aggregated payoff of node i is given by the sum of the payoffs π_i over all layers. Furthermore, after each round of the game, individuals update their strategies with a rule that can use the global knowledge about the benefits of the neighbors or be random [138].

Finally, it worth noting that cooperation is also studied in public good games, where the prisoner's dilemma is played in overlapping groups of individuals: cooperators contribute with a "cost" d to the public good, while defectors do not contribute and the total amount in the common pool of each group is multiplied by a factor r and distributed equally among all members of the group [248].

Previous research [139] demonstrated that the resilience of cooperative behavior can be enhanced by the multiplex structure through the simultaneous formation of correlated clusters of cooperators across different layers – that is, through multiplex network reciprocity. However, it was also proven [37] that, to gain benefits from multiplex structure, a high topological overlap is needed or, in other words, individuals have to be similarly linked across different layers (see Figure 38, panels [a] and [b]). Other studies [163] investigated the role of degree correlation v between nodes in different layers for the emergence of cooperation. They found that in the absence of degree correlations, increasing the number of layers only leads to mild changes. However, if degree correlations are present, we observe a mean final cooperation of $c = 0.5$, and this value is nearly independent of the game payoff parameters. This mechanism is called *topological enslavement* and can be understood by considering that, if degree correlations are strong, hubs dominate the game dynamics, since they have the potential to earn higher payoffs (because they play more games) and they are more likely to be selected by other nodes as imitation candidates.

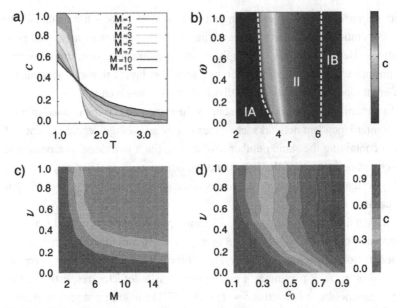

Figure 38 (a) Mean final cooperation as function of the temptation to defect T, for several multiplex networks with different numbers of layers M. (b) Multiplex with $M = 2$ layers, where cooperation is studied as a function of synergy factor r and edge overlap ω. Dashed lines separate regions with either full defection $c = 0$ (IA) or full cooperation $c = 1$ (IB), from the region of continuously evolving coexistence of cooperators and defectors $0 < c < 1$ (II). Panel (a) readapted from [139] and panel (b) readapted from [37]. (c) Mean final cooperation (color coded) for the prisoner's dilemma as a function M and the strength of degree correlations ν. (d) Mean final cooperation as a function of the initial density of cooperators c_0 and the strength of correlations ν. Panels (c) and (d) readapted from [163].

Furthermore, topological enslavement implies that the outcome of the evolution of the system is determined by the initial conditions (see Figure 38, panels [c] and [d]).

Finally, as anticipated in Section 3.5, layer-layer correlations might have a deep impact on the dynamics on the top of multilayer systems. This is the case for game dynamics, where cooperation might be hampered, rather than enhanced, by specific correlation patterns combining assortative and disassortative degree mixing across layers [111, 293] (see Figure 39 for details).

4.4 Interdependent Processes

The second class of dynamical processes on multilayer networks is the one of "interdependent" or "coupled" dynamics, consisting of systems characterized by *different* processes on each layer. The inter-layer interactions couple

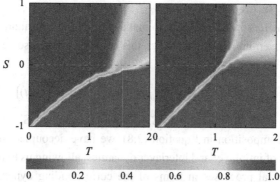

Figure 39 Fraction of cooperators as a function of the payoff of a cooperator when playing with a defector (*S*) and the payoff of a defector when facing a cooperator (*T*). The interaction layer is subject to disassortative mixing (assortative coefficient $A_I < 0$) while the updating network is subject to assortative mixing (disassortative coefficient $A_U > 0$). Results show the disruptive effect of symmetry breaking for the evolution of cooperation. Parameter values are $A_I = -0.1$ and $A_U = 0.1$ (Left) and $A_I = -0.2$ and $A_U = 0.2$ (Right). Figure from [293].

such processes and are responsible for emerging phenomena that could not be detected in a single-layer network framework (see Table 3).

As in the case of the structure, it is possible to introduce a unifying framework in terms of a dynamical SNXI decomposition, outlined in Equation (2.8), to describe dynamics on multilayer networks. Let $x_{i\alpha}^{[l]}$ (where $l \in 1, 2, \ldots, C$) denote the *l*-th component of a $C - dimensional$ vector $\mathbf{x}_{i\alpha}$ that represents the state of node i in layer α. The most general (and possibly nonlinear) dynamics governing the evolution of each state is given by the systems of equations

$$\dot{\mathbf{x}}_{i\alpha}(t) = F_{i\alpha}(\mathbf{X}(t)) = \sum_{\beta=1}^{L} \sum_{j=1}^{N} f_{i\alpha}^{j\beta}(\mathbf{X}(t)) \tag{4.28}$$

$$= \underbrace{\sum_{\beta=1}^{L} \sum_{j=1}^{N} f_{i\alpha}^{j\beta}(\mathbf{X}(t)) \delta_{\alpha}^{\beta} \delta_{i}^{j} + \sum_{\beta=1}^{L} \sum_{j=1}^{N} f_{i\alpha}^{j\beta}(\mathbf{X}(t)) \delta_{\alpha}^{\beta} \left(1 - \delta_{i}^{j}\right)}_{\text{intralayer dynamics}}$$

$$+ \underbrace{\sum_{\beta=1}^{L} \sum_{j=1}^{N} f_{i\alpha}^{j\beta}(\mathbf{X}(t)) \left(1 - \delta_{\alpha}^{\beta}\right) \delta_{i}^{j} + \sum_{\beta=1}^{L} \sum_{j=1}^{N} f_{i\alpha}^{j\beta}(\mathbf{X}(t)) \left(1 - \delta_{\alpha}^{\beta}\right) \left(1 - \delta_{i}^{j}\right)}_{\text{interlayer dynamics}}$$

$$= \underbrace{f_{i\alpha}^{i\alpha}(\mathbf{X}(t))}_{\text{self-interaction}} + \underbrace{\sum_{j \neq i} f_{i\alpha}^{j\alpha}(\mathbf{X}(t))}_{\text{endogenous interaction}} + \underbrace{\sum_{\beta \neq \alpha} \sum_{j \neq i} f_{i\alpha}^{j\beta}(\mathbf{X}(t))}_{\text{exogenous interaction}} + \underbrace{\sum_{\beta \neq \alpha} f_{i\alpha}^{i\beta}(\mathbf{X}(t))}_{\text{intertwining}},$$

$$= \mathbb{S}_{i\alpha}(\mathbf{X}(t)) + \mathbb{N}_{i\alpha}(\mathbf{X}(t)) + \mathbb{X}_{i\alpha}(\mathbf{X}(t)) + \mathbb{I}_{i\alpha}(\mathbf{X}(t))$$

where $\mathbf{X}(t) \equiv (\mathbf{x}_{11}, \mathbf{x}_{21}, \ldots, \mathbf{x}_{N1}, \mathbf{x}_{12}, \mathbf{x}_{22}, \ldots, \mathbf{x}_{N2}, \ldots, \mathbf{x}_{1L}, \mathbf{x}_{2L}, \ldots, \mathbf{x}_{NL})$ and we did not use the tensorial formalism to make explicit the contributions of each term in the governing equation. In fact, this equation could also be compactly written as

$$\dot{x}_{j\beta}(t) = F[x_{j\beta}(t)] = \mathbb{S}[x_{j\beta}(t)] + \mathbb{N}[x_{j\beta}(t)] + \mathbb{X}[x_{j\beta}(t)] + \mathbb{I}[x_{j\beta}(t)]. \tag{4.29}$$

We call this equation "dynamical SNXI decomposition." Similarly to the structural decomposition in Equation (2.8), we have decoupled the different contributions of intralayer and interlayer dynamics, allowing us to classify different dynamical processes in terms of the corresponding dynamical SNXI components. The peculiar behavior of the interdependent processes is ascribed to the exogenous and intertwining components. Although the most-studied examples come from mixed spreading processes, which are crucial for understanding phenomena such as the spreading dynamics of two concurrent diseases in two-layer multiplex networks [81, 89, 109, 243, 249], and the spread of diseases coupled with the spread of information or behavior [122, 123, 141, 142, 182, 294], other types of dynamical interdependence are attracting a growing interest.

Coupling Diffusion with Synchronization

An illustrative and pedagogical example has been proposed by [208]. In this work, the authors examine the interdependent dynamics of the two processes presented in the two previous sections, namely diffusion and synchronization. They propose a model that mimics the interplay between the neural activity and energy transport in brain regions, from which a rich collection of behaviors emerges. These processes evolve in the two layers of a multilayer network and are related to each other by the correspondence between layers, which in this case is realized through the functional relation between the parameters governing the two processes and the state variables. In particular, the dynamics of the entire system is governed by the following equations:

$$\begin{cases} \dot{x}_i = F_{\omega_i}\left(\mathbf{x}, A^{[1]}\right) \\ \dot{y}_i = G_{\chi_i}\left(\mathbf{y}, A^{[2]}\right) \end{cases} \quad i = 1, 2, \ldots, N, \tag{4.30}$$

where $\mathbf{x} = \{x_1, x_2, \ldots, x_N\} \in \mathbb{R}^N$ and $\mathbf{y} = \{y_1, y_2, \ldots, y_N\} \in \mathbb{R}^N$ denote the states of the two dynamical processes, while the topologies of the two layers are encoded in the adjacency matrices $A^{[1]} = \left\{a_{ij}^{[1]}\right\}$ and $A^{[2]} = \left\{a_{ij}^{[2]}\right\}$, respectively, such that $a_{ij}^{[1]} = 1 \left(a_{ij}^{[2]} = 1\right)$ if a link exists between nodes i and j in the first (second) layer, and $a_{ij}^{[1]} = 0 \left(a_{ij}^{[2]} = 0\right)$ otherwise. The evolution of the system depends on a set of parameters ω and χ, which in turn depend on the state of the nodes in the adjacent layer:

$$\begin{aligned}\dot{\omega}_i &= f(\omega_i, y_i) \\ \dot{\chi}_i &= g(\chi_i, x_i)\end{aligned} \quad i = 1, 2, \ldots N. \tag{4.31}$$

The authors assign to functions F_{ω_i} and G_{χ_i} a Kuramoto dynamic and a continuous-time random walk, respectively. Subsequently, they assign the functions f and g in Equations (4.31), respectively, relating the frequency ω_i of the oscillator i at layer 1 to the state y_i at layer 2, and the bias property χ_i of the random walkers at layer 2 to the oscillator phase x_i at layer 1. The natural frequency ω_i of the oscillator i evolves, relaxing to values proportional to the fraction of random walkers at the replica node i in the other layer. Analogously, the random walks are biased toward (away from) strongly synchronized nodes.

Note that the coupling between the two layers is tunable through two parameters λ and α that represent, respectively, the intensity of the influence of the random walk on the oscillators and vice versa. This setup completely defines an interdependent process on a multilayer network: depending on the coupling strengths λ and α, the collective behavior exhibits special dynamics (see Figure 40). The random walkers are homogeneously distributed in the incoherent state, while in the synchronized state, the distribution is heterogeneous. Conversely, at certain values of the tuning parameters, the system encounters an ES, or a bistability region characterized by a hysteretic behavior. Similar phase transitions can also be observed in single-layer networks under certain conditions, as reported in [2, 67, 186]. Importantly, the multilayer network model offers a parsimonious explanation of the emergence of these collective phenomena, considering explicitly the intertwined nature of the dynamics.

Coupling Epidemics Spreading with Awareness Diffusion

Many different phenomena in nature can be described, in their essence, by the results of constructive or destructive relations between two or numerous parts. For instance, the mutualistic or competitive relation between dynamical systems gives rise to a wealth of fascinating behaviors.

The dynamics occurring in one layer can have positive or negative feedbacks on another: for example, human behavior (e.g., information awareness) can inhibit the spread of disease; social mixing between classes and mobility may produce abrupt changes in the critical properties of the epidemic onset; cooperation emerges where the classical expectation was defection.

A generalization of these results can be found in [88], in which the authors describe some universal features of interdependent systems with coupled

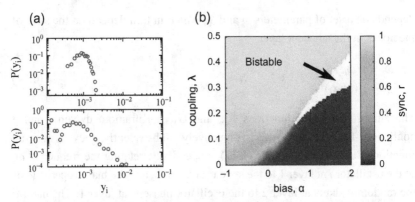

Figure 40 Coupling diffusion with synchronization dynamics. (a) Distribution $P(y_i)$ of steady-state random walker fractions y_i at layer 2 for $\alpha = 1.0$, when the oscillators at layer 1 are incoherent ($\lambda = 0.1$, top) and synchronized ($\lambda = 0.4$ bottom). (b) Synchronization phase diagram showing the level of synchronization as a function of coupling λ and bias exponent α. The bistable region is colored in white. Figure adapted from [208].

dynamics. There is an interesting relation between the collective behavior emerging from percolation processes and from the ones arising in interdependent systems. In such processes, abrupt transitions and critical dynamics may arise in certain circumstances and, in dynamically coupled systems, other interesting phenomena may also be observed, such as hysteresis and multistability, with functionality of nodes in one layer influencing the functionality of their replicas in the other layers. In [88], the functionality is quantified by an order parameter, which plays a role in the coupling strength between the nodes in different layers, with the order of a node dynamics affecting the order of its neighbors. This fact is the dynamical counterpart of interdependent percolation, where the functionality of a node is related to its belonging to the mutual giant connected component (see Section 4.5 for details). It turns out that the dynamic interdependence increases the vulnerability of the system [87], as in the case of percolation processes.

Epidemic models are another important example of systems in which interdependencies play a crucial role (see [103, 294] for a review) and, also in this case, hysteretic behavior with abrupt transitions may occur. This means that, for instance, explosive pandemics with no early warning can suddenly appear and, in a similar implosive way, can disappear. Unfortunately, the value of the infection rate that can eradicate the disease must be much lower than the value that triggered it [88]. Other interesting behaviors arise in the case of competitive diseases, where mutual exclusion or endemic coexistence may spontaneously occur.

Figure 41 shows two phase diagrams of disease incidence of reciprocally enhanced (right) and inhibited (left) disease-spreading processes. In the figure is highlighted the existence of a curve of critical points that separate endemic and non-endemic phases of the disease. Moreover, the curve that separates the endemic and non-endemic phases, is in turn divided into two parts: one in which the critical properties of one spreading process are independent of the other (straight dashed line), and one in which the critical properties of one spreading process do depend on those of the other layer (solid curve). The point at which this crossover occurs is called a *metacritical point*.

In fact, traditional epidemic models can describe in detail the spreading of a single disease in different realistic situations, whereas the multilayer representation can extend the possibilities to spreading processes with interacting diseases. In [249], the authors propose a framework to describe the spreading dynamics of two concurrent diseases (see Figure 42). Using susceptible-infected-susceptible-susceptible-infected-susceptible (SIS-SIS) and susceptible-infected-recovered-susceptible-infected-recovered (SIR-SIR) models with appropriate interlayer couplings describing the influence of one disease on the other, they derive the epidemic threshold for the two interacting diseases, showing that the onset of a disease's outbreak is conditioned to the prevalence levels of the other disease.

Epidemic thresholds can be influenced not only by the presence of a second disease but also by the individual's change of awareness about an ongoing epidemic. In [141], a multilayer network couples the dynamics of disease

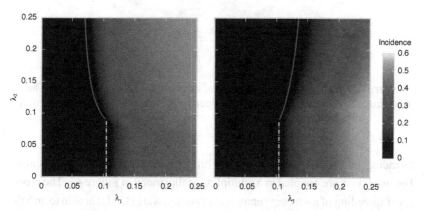

Figure 41 Two (left) reciprocally enhanced and (right) reciprocally inhibited disease-spreading processes of susceptible-infected-susceptible (SIS) type. The colors in the figure represent the prevalence levels of the diseases at a steady state of Monte Carlo simulations. Note the emergence of a curve of critical points (at a "metacritical point") in which the spreading in one layer depends on the spreading in the other. Figure from [103].

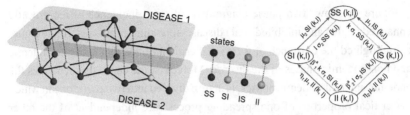

Figure 42 An SIS-SIS interacting diseases model. Left: multiplex representation of the diseases spreading. Each individual is present in both the layers and can be infected by one (or both) of the diseases, as indicated in the central panel. Right: possible transitions between the different states of the SIS model. The variables represent the densities of individuals of each type having degree equal to k in the first layer and degree l in the second. Figure from [249].

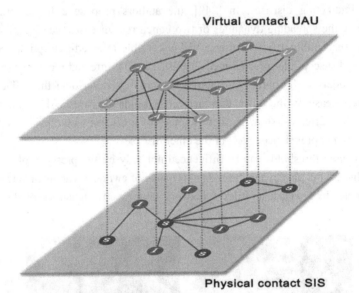

Figure 43 Multiplex representation of awareness-disease spreading. The spreading of awareness occurs in the upper layer, whereas the spreading of the disease takes place in the lower layer. Figure from [141].

spreading in a social network and the diffusion of awareness among actors. The two layers are coupled in a multiplex, as illustrated in Figure 43. The process of spreading of awareness unaware-aware-unaware (UAU) is akin to an SIS process, where in place of susceptible (S) there are unaware (U) and in place of infected (I) there are aware (A) actors. The probability of being infected is influenced by the state of awareness of the individuals. The authors found the existence of a *metacritical* point in which the state of awareness of the individuals can control the onset of an epidemic.

Coupling Game Dynamics with Opinion Dynamics

The multilayer representation of the dynamics of complex systems can also be used to shed light on real-world social dilemmas. The influence of factors acting on different layers might explain the emergence of particular patterns of cooperation between social agents. In [10], the authors present a model coupling evolutionary game dynamics and opinion dynamics, regarded as two processes evolving on distinct layers of a multiplex network.

To model the game dynamics on the first layer, the authors adopt a *replicator* in which the individuals (nodes) copy the strategy of one of their neighbors with a probability that depends on the payoff difference, assigning higher probabilities to the strategies of players that have earned a higher payoff compared to copying one's node The possible states are "cooperate" and "defeat." The opinion dynamics on the second layer is modeled using the *voter model*, where individuals (nodes) adopt the opinion of a randomly selected neighbor with a certain probability. The opinions of the nodes can be "cooperate" or "defeat," as in the game layer. The authors assume that imitating a cooperative opinion is more likely than imitating a defection. This can be interpreted as the influence of media campaigns or broadcasting agents. Depending on the specific parametrization of both the game and the opinion dynamics, this model gives rise to fascinating dynamical behaviors in which equilibria of different types exist for the game dynamics. The impact of social influence on the decision of individuals is conveyed by the interlayer coupling, which is encoded in the parameter γ, representing the tendency of individuals to act in agreement with their proclamations: the nodes in one layer copy their own state from the other layer with probability γ.

The main result is that this model is sufficient to explain the emergence of cooperation in scenarios where the pure game dynamics predicts defection. This is due both to the intertwined dynamics and to the multilayer structure itself. In fact, the authors proved that the geometric correlation between layers has a significant impact on the stability of the system. Importantly, under certain conditions of correlation between layers, the system can reach a polarized metastable state (see Figure 44). This result can explain the observed polarization in real-world social systems.

Ultimately, the emergence of cooperation in unexpected conditions is due to the interplay between the coupled dynamics of strategies and opinions, the complex topologies of the networks upon which the dynamics exert, and the structural relations between the layers. Missing one of these elements could hinder the right interpretation of the complex behavior of such systems.

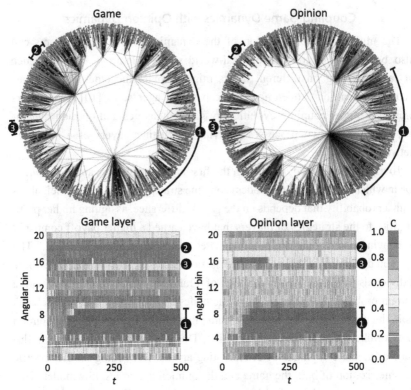

Figure 44 Polarization of opinions and strategies in a multiplex with 5,000 nodes, in the presence of angular correlations between the layers. Top: comparison between the two layers; the color is the time average of the state of each node. Bottom: evolution of the density of cooperators in the angular bins used to compute the interlayer correlations. Numeric labels indicate the clusters of nodes that adopt the same strategy. Figure from [10].

Coupling Epidemics Spreading with Social Integration

A multilayer perspective on epidemic spreading can be a valuable support to design more educated strategies to reduce disease risk. A recent example comes from [59], which integrates social dynamics, human mobility, and epidemic spreading to assess the risk of measles outbreak in Turkey. During the past decade, Turkey has received more than 3.5 million refugees coming from Syria. The levels of immunization of the two populations are considerably different. The outbreak risk is analyzed through a multilayer transmission model, which takes into account the different levels of immunization in the two populations, along with the mobility pattern and the level of social integration. The main result of the study is that, in the case of heterogeneous immunization, high levels of social interaction can drastically reduce the spatial spread and incidence of a disease. This apparently counterintuitive result is due to the fact that the high immunization coverage of one population (Turkish citizens) can shield the other (Syrian refugees) from getting exposed to the infection as an effect of herd immunity. The network structure is defined by dividing Turkey into patches corresponding to administrative areas that represent the nodes. The layers of the network are encoded by Turkish (T) and Syrian refugee (R) populations. On this network, social dynamics, mobility, and epidemic spreading happen simultaneously.

The authors denote with $c_{ki}^{(p)} (p \in T, R)$ the elements of a matrix $\mathbf{C}^{(p)}$ encoding the number of people belonging to population $p \in \{T, R\}$ traveling from patch k to patch i, and with α the fraction of Syrian contacts with Turkish citizens. The force of infection for each population in the i-th patch depends on the contribution of all patches in the country:

$$\lambda_i^{(T)}\left(\alpha, \mathbf{C}^{(T)}, \mathbf{C}^{(R)}\right) = \beta_T \sum_{k=1}^{L} \left[\underbrace{c_{ki}^{(T)} \frac{I_k^{(T)}}{N_k^{(T)}}}_{\text{Endogenous}} + \underbrace{\alpha c_{ki}^{(R)} \frac{I_k^{(R)}}{N_k^{(R)}}}_{\text{Exogenous}} \right],$$

$$\lambda_i^{(R)}\left(\alpha, \mathbf{C}^{(T)}, \mathbf{C}^{(R)}\right) = \beta_R \sum_{k=1}^{L} \left[\underbrace{\alpha c_{ki}^{(T)} \frac{I_k^{(T)}}{N_k^{(T)}}}_{\text{Exogenous}} + \underbrace{c_{ki}^{(R)} \frac{I_k^{(R)}}{N_k^{(R)}}}_{\text{Endogenous}} \right]$$

where $\beta_p = \beta / P_i^{(p)}(\alpha, c)$ is the transmission rate for population p and $P_i^{(p)}(\alpha, c)$ is an appropriate normalization factor. From this equation we can easily recognize the structure of Equation (4.28), where the contributions to the force of infection come from an endogenous term, accounting for the infectivity due to individuals from the same population, and an exogenous term, accounting for the infectivity due to the other population. The parameter α is the level of social mixing and can be changed according to different social integration scenarios. This is the parameter that plays the role of the coupling between the layers. An illustrative representation of the model is shown in Figure 45.

The epidemic transmission dynamics is eventually regulated by the following SIR dynamical model

$$\begin{cases} \dot{S}_i^1 = -\lambda_i^1(\alpha, c) S_i^1 \\ \dot{S}_i^2 = -\lambda_i^2(\alpha, c) S_i^2 \\ \dot{I}_i^1 = \lambda_i^1(\alpha, c) S_i^1 - \gamma I_i^1 \\ \dot{I}_i^2 = \lambda_i^2(\alpha, c) S_i^2 - \gamma I_i^2 \\ \dot{R}_i^1 = \gamma I_i^1 \\ \dot{R}_i^2 = \gamma I_i^2 \end{cases}$$

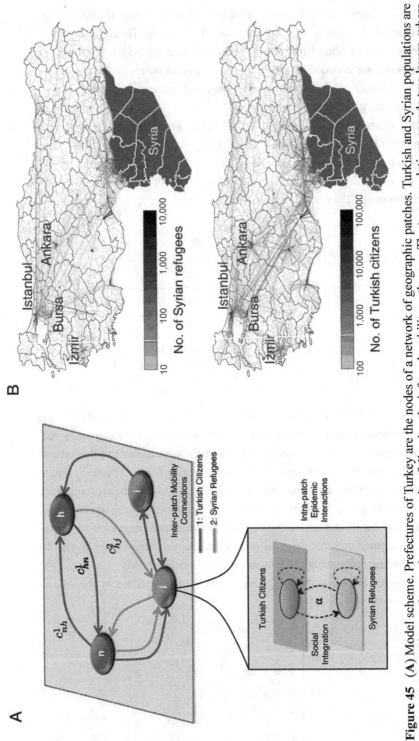

Figure 45 (**A**) Model scheme. Prefectures of Turkey are the nodes of a network of geographic patches. Turkish and Syrian populations are encoded by two colors and move between patches following the inferred mobility pathways. The two populations encode two layers, where social dynamics and epidemic spreading happen simultaneously. (**B**) Mobility of Syrian refugees (upper) and Turkish citizens (lower) inferred from mobile phone data. Figure from [59].

whose analysis suggests that the incidence of the measles can be reduced up to 90% in the case of very high levels of integration [59].

Despite the different contexts, the multilayer dynamics can have positive or negative feedbacks, leading to interdependence between the corresponding critical points of the dynamics. As a consequence, two different regimes exist: (i) one in which the critical properties of one process depend on those of the other, and (ii) one in which the critical properties are independent of the other. The two regimes are separated by a metacritical point, where a crossover occurs (see a recent review from [103] ad Figure 41). This regime shift is analogous to the one occurring in percolation processes, which is presented in the next section.

4.5 Percolation

In this section, we switch our attention to percolation [268], where the goal is to analyze how different properties of the network change as we remove some of its nodes or links. The properties we are interested in are typically topological, such as the size of the largest connected component or the distribution of small connected components, and they are used as a first proxy to assess the functionality of a system exposed to failures or attacks. Failures are modeled as random removals, whereas attacks assume some a priori knowledge of the network, where the elements are ranked given some criterion – frequently topological (degree, betweenness, etc.) [7, 80], albeit not strictly necessary [20, 22] – and removed accordingly. We aim at presenting the basic mathematical framework to address the problem of multilayer percolation and showcase some applications. Good reviews to expand on what we present here can be found in [51, 176].

The first step toward a description of multilayer percolation processes is to consider the generalization of the generating function methodology, frequent in single-layer networks [204]. In [177], random percolation is studied in general multilayer networks described by the set of degree distributions $\{p^{\alpha}_{k_1 k_2 \ldots k_L}\}$, where α is the layer label and k_β denotes the number of links toward nodes in layer β. Expressions for the size of the giant component of each layer S^α can be written as a function of the generating functions. This framework allows us to consider percolation on correlated multilayer networks, leading to nontrivial results. For example, in [175], correlated duplexes of Erdős-Rényi networks are studied. When degrees across layers are maximally anti-correlated – that is, hubs in one layer are low-degree nodes in the other layer – the giant component appears at a link density considerably higher than the value for which it appears in uncorrelated duplexes. The giant component exists, though, for

any nonzero link density if interlayer degrees are maximally positively correlated. In other words, the more correlated the degrees are, the fewer edges are
needed to make a macroscopic structure emerge. When it comes to attacks on
the most connected nodes, the latter statement is true up to a certain intralayer
mean degree (assuming it is the same for both layers), after which the behavior
is reversed and maximally negative correlated networks become more robust
[193].

This framework disregards any functional characteristics of the layers and,
from a phenomenological point of view, continuous phase transitions are
always found. This is no longer true if the nature of the layers is taken into
account, for example via the interdependence of the nodes or antagonistic interactions. These more realistic scenarios define new conditions for a node to
remain in the network and need alternative, more function-oriented metrics to
describe the state of the system. A widely accepted choice is the mutually connected component [68, 263] already defined in Section 3.2, which is the set
of nodes that are connected within each and every layer simultaneously. The
most striking result is that when the giant mutual component (GMC) is computed in interdependent networks, the percolation phase transition changes its
nature, becoming a discontinuous one [263]. This has serious implications for
the robustness of the system since the disintegration occurs abruptly – that is,
it is hard to anticipate. Mathematically, the idea is as follows.

Let us assume an edge-colored multigraph, let $p_{k_1...k_L}$ be the probability that
a node has degree k_α to other nodes within the layer α, and let $q_{k_1...k_L}$ be the
corresponding excess degree distribution. We indicate by w_α the probability
that a node does not belong to the GMC via a link in layer α. Hence, $w_\alpha^{k_\alpha}$
gives the probability that the node does not belong to the GMC via any of its
neighbors in layer α. The condition to belong to the GMC is that the node has to
be connected to it in all the layers – that is, the size of the GMC is proportional
to $\prod_{\alpha=1}^{L}(1 - w_\alpha^{k_\alpha})$. We just need to rescale by the occupation probability ϕ and
average over the degree distribution, yielding

$$M = \phi \sum_{k_1=0}^{\infty} \cdots \sum_{k_L=0}^{\infty} p_{k_1...k_L} \prod_{\alpha=1}^{L}(1 - w_\alpha^{k_\alpha}). \tag{4.32}$$

To compute w_α, we first note that $1 - w_\alpha$ is the probability that a node at the
end of a link in layer α belongs to the GMC. For this to happen, any of its
remaining $k - 1$ neighbors in layer α are in the GMC as well. Moreover, due
to the condition of mutual connectivity, in every other layer, the node needs to
belong to the GMC via any of its neighbors. Rescaling by ϕ because the node
needs to be present in the network, and averaging, we obtain

$$1 - w_\alpha = \phi \sum_{k_1=0}^{\infty} \cdots \sum_{k_L=0}^{\infty} q_{k_1 \ldots k_L} \left(1 - w_\alpha^{k_\alpha - 1}\right) \prod_{\substack{\beta=1 \\ \beta \neq \alpha}}^{L} (1 - w_\beta^{k_\beta}). \qquad (4.33)$$

By inserting the solutions $\{w_\alpha\}$ of this system of equations into Equation (4.32), we readily obtain the size of the GMC. See Figure 46 to appreciate the emergence of the abrupt transition of the GMC for multiplex systems with Erdős-Rényi networks in the layers. The discontinuity is also present when considering multiplexes of scale-free networks, but, at odds with single-layer, scale-free percolation, the occupation probability ϕ_c is finite, making them more vulnerable to random failures than the single-layer network [39]. Targeted interventions in the network can be easily modeled as well by including a degree-dependent occupation probability $\phi_{k_1 \ldots k_L}$ inside the sums [193, 307].

The dependence of the mutual component M shown in Figure 46 for Erdős-Rényi multiplexes is shared for other intralayer topologies as well. That is, when the number of layers increases, the value of the mean degree at which the discontinuity appears becomes larger, thus broadening the parameter region where the network is not functional. Moreover, the height of the discontinuity jump becomes larger too, thus making the transition harder to anticipate when going from the supercritical to the subcritical region. In light of these results, the more layers, the more fragile the system is, which might seem paradoxical from an evolutionary point of view: why would a system organize itself in a layered structure if that reduces its robustness? In [232], Radicchi and Bianconi provided a potential answer to this conundrum by proposing a model of multilayer percolation that relaxes the condition for functionality of the nodes. They argue that a node can be functional – that is, it is not removed, as far as it is functioning in at least a pair of layers. This new condition for functionality allows them to conclude that the addition of extra layers boosts the robustness of the system.

The abrupt nature of the transition induced by Equation (4.33) holds as far as $L > 1$; see Figure 46. When $L = 1$, the mutual component coincides with the standard giant component, so the transition is continuous. We cannot interpolate continuously from $L = 1$ to $L = 2$ to understand how the nature of the phase transition changes. However, there are other variables that we can tune to go from an effective single-layer system to a multiplex.

The first of these variables is the multiplexity parameter. It might occur that in real multiplex networks, only a fraction q of all nodes shares the functional dependency across layers, hence a natural question is what is the nature of the transition as a function of q. Does it suffice to have a nonzero fraction of dependent nodes to observe the discontinuity, or, on the contrary, is there a

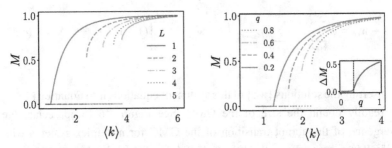

Figure 46 Percolation phase transition in a multiplex formed by intralayer Erdős-Rényi networks, with the same mean degree $\langle k \rangle$. On the left, we display M, which is the solution of $M = \phi(1 - e^{-\langle k \rangle M})^L$, as a function of the mean degree, for different number of layers, with the occupation probability fixed to $\phi = 1$. On the right, we study the emergence of the giant mutual component for partial multiplexes, where q is the fraction of interdependent nodes. In this case, M is the solution of $M = \phi(1 - e^{-\langle k \rangle M})(1 - qe^{-\langle k \rangle M})$. In the inset, we show the height of the jump of the order parameter at the transition point, along with the theoretical value of the tricritical point (vertical dashed line).

finite threshold only above which we observe an abrupt transition? To answer this question, let us focus on duplexes. If a node in one of the layers has a dependency link, which occurs with probability q, then the condition to belong to the GMC is the same as discussed earlier. Instead, if the node does not have a dependency link, which occurs with probability $1 - q$, then it belongs to the GMC as far as any of its neighbors of its very same layer belong to the component. Therefore, we can write

$$M_\alpha = q\phi \sum_{k_1=0}^{\infty} \cdots \sum_{k_L=0}^{\infty} p_{k_1 \ldots k_L} \prod_{\beta=1}^{L}(1 - w_\beta^{k_\beta})$$

$$+ (1 - q)\phi \sum_{k_1=0}^{\infty} \cdots \sum_{k_L=0}^{\infty} p_{k_1 \ldots k_L} w_\alpha^{k_\alpha}. \tag{4.34}$$

The set of probabilities $\{w_\alpha\}$ obeys the equations

$$1 - w_\alpha = q\phi \sum_{k_1=0}^{\infty} \cdots \sum_{k_L=0}^{\infty} q_{k_1 \ldots k_L} \left(1 - w_\alpha^{k_\alpha - 1}\right) \prod_{\substack{\beta=1 \\ \beta \neq \alpha}}^{L}(1 - w_\beta^{k_\beta})$$

$$+ (1 - q)\phi \sum_{k_1=0}^{\infty} \cdots \sum_{k_L=0}^{\infty} q_{k_1 \ldots k_L}(1 - w_\alpha^{k_\alpha - 1}). \tag{4.35}$$

Focusing again on a duplex of Erdős-Rényi networks, we see in Figure 46 that there is a finite tricritical point q_c at which the transition changes its order, confirming that, in general, a multiplex needs a certain level of interdependency between layers to experience the discontinuous transition. Since many infrastructural networks are embedded in a two-dimensional space, in a similar fashion, one might ask what type of transition we encounter when coupling low-dimensional networks, which alone show continuous transitions, in a multiplex. It turns out that in this case the transition does not change its order [45, 262].

The second variable that allows us to interpolate between multiplexes and monoplexes is the link overlap across layers. For complete overlapping, all layers are equal and the problem is reduced to percolation of a single layer, showing a continuous transition. How much overlap do we need to observe the abrupt transition? Different approaches and techniques have been used to answer this question, and accordingly slightly different phenomenology has been discovered. Although the details of the phase diagram depend on the number of layers and the degree distributions, Hu and coauthors have found that the transition stays abrupt as far as we do not have complete overlapping [150]. In [74], it has been reported that a critical value of the edge overlap exists that changes the nature of the phase transition from a hybrid first-order to a continuous one. See [75] for the generalization of the theory to an arbitrary number of layers. If the edge overlap is combined with other topological correlations, the phase diagram displays multiple and recursive hybrid phase transitions [40]. In [194], the problem of link overlap has been formulated in a way that the discontinuous transitions display hysteresis. The role of the edge overlap, together with other topological correlations, has been also explored in the problem of multilayer optimal percolation [214] – that is, in the identification of the smallest set of nodes that, when removed, cause the largest damage in the network [245]. This problem is known to be NP-hard in single-layer networks [159], and although there is no equivalent proof in multilayer architectures, the intuition indicates that it is so as well. On this basis, heuristic approaches to identify the critical subset of nodes predominate. In [245], the efficiency of 20 of these strategies is evaluated. The authors find that when no structural correlations exist, a family of Pareto-efficient strategies based on both structural descriptors and multi-objective optimization is the best at dismantling the network. However, when evaluated in real multilayer networks that present nontrivial correlations, the variability in performance changes from one dismantling strategy to another, suggesting that a fair assessment of multilayer robustness requires a comparison between strategies.

Beyond node interdependency as a condition for functioning, other interesting mechanisms have been considered in the literature. One of them is antagonistic relations, where a node in one layer is functional – that is, it has not been removed – only if its replicas are not [308]. Examples of this kind of competitive or noncooperative relations might be relevant, for example, in biological networks. Interestingly, it has been found that when percolation is considered in this scenario, the abrupt transition shown in what follows persists, but displays as well, at variance with the transition of the mutual component introduced earlier, the typical hysteresis and bistability behaviors of equilibrium thermal first-order phase transitions [135]. Other mechanisms have followed to include these behaviors too – for example, in [192, 194].

Note that all of these results derived with the machinery of generating functions come with some underlying assumptions that are very rarely met when studying real networks: (i) the network is tree-like, (ii) it has infinite size, and (iii) the quantities of interest are computed as an average over the ensemble of networks with given degree distribution, although, in reality, we only have access to one supra-adjacency matrix, among all the possible ones that a graph model could generate. As a consequence, the analytical predictions might deviate from the actual process of percolation, obtained, for example, via simulations. Analytical approaches have been proposed to overcome points (ii) and (iii) in [52, 230], where a percolation theory of multiplex and interdependent networks is introduced that takes as input the adjacency matrix instead of the degree distribution. Furthermore, in a real network, we may need to assess the robustness under perturbation scenarios related to node metadata. For example, in [28], it is addressed, among other things, how a multiplex network capturing the flow of subsistence-related goods and services among households in several Alaskan Native communities responds to perturbations involving targeted removals of specific resources by category – for example, terrestrial, marine, or riverine, as a representation of natural disasters. An analytical framework to take into account non-topological features in the robustness of a network has been recently developed for single layers [20], but, at present, there is not a generalization for multilayer networks.

4.6 Cascade Failures

The percolation model has the limitation that failures and attacks are treated statically. A more complete description of multilayer robustness and resilience would include their time evolution, in order to better understand under which conditions small perturbations can trigger global network-wide effects, the so-called *cascades*. In this section, we review some of the most emblematic models for multilayer cascade failures.

The seminal work of Buldyrev and colleagues was one of the first to deal with the propagation of failures across interdependent multiplexes [68]. The motivation for such a study was the 2003 Italy blackout, for which it was hypothesized that the power grid and the internet network, the latter acting as a supervisory control and data acquisition system, were interdependent, and that failures in the power stations hampered the internet communication and further propagated the malfunction across the system (see the top panels of Figure 47).

However, the interest in the amplification of small perturbations throughout the network is much more general, finding eventual applications in areas such as biochemistry, where metabolic pathways interact in complex ways with key tissues, or finance, where different banks might share the same asset in their balance sheet [153], among many other examples. In [68] was introduced the concept of a mutually connected component from a dynamical perspective, at odds with its static definition presented in the previous section (see the bottom panels of Figure 47). It was shown that these cascades evolving in multiplex networks yield a discontinuous phase transition in the size of the GMC [68], meaning that the failure of one single node from an apparently healthy, functional infrastructure can generate the emergence of a global cascade that collapses the entire system. It was later shown that these cascading failures can be mapped to static percolation in multiplexes [263]. In fact, many of the results presented in Section 4.5 are recovered at the final state of the cascade propagation of this type of model.

Despite the fact that the dynamical propagation of cascades needs a more convoluted mathematical treatment than percolation, these techniques have been very flexible at adapting to variations of the original work of Buldyrev, so that the propagation of cascades is modeled in more realistic scenarios. We briefly review some of them in the following, and refer the interested reader to the more complete reviews [160, 253, 284].

After [68], the problem of dependency-based cascades has been generalized to L layers in [129], allowing analytical, closed solutions in certain interdependent setups. The problem of failure propagation in multiplexes with partial interdependency, where not all nodes have dependency connections, has been addressed in [218]. If the fraction of interdependent nodes is decreased enough, the transition is no longer abrupt but becomes second-order. In order to narrow the range of parameters for which the abrupt transition occurs, different strategies to choose the autonomous nodes, those without interdependencies, have been proposed [250]. It turns out that selecting those with the highest degree or highest betweenness significantly reduces the likelihood of an abrupt collapse [250]. Intralayer correlations can be included in the analysis as well. In [152], interdependent multiplexes with a tunable average number of single links and an average number of triangles per node are analyzed, finding that, for fixed average degree, the higher clustered networks are less robust than those with lower clustering. Intralayer correlations might be coupled as well

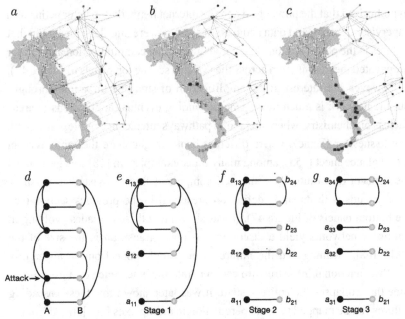

Figure 47 Top: evolution of a cascade of failures in a real interdependent system composed by a power network (on the map) and an internet network (shifted above) and involved in the Italian blackout of September 2003. In (a), the failure of a power station makes the Internet nodes to stop functioning, indicated in the map. In the internet network it is indicated the nodes that disconnect from the largest connected component in the next cascade step. These nodes will fail immediately after (b), introducing a feedback of failures back in the power network. Those nodes that have been isolated in (b) are the ones that will fail in the next event (c), inducing further interdependent failures in the Internet network. Bottom: we sketch this process for further clarification. In **d**, an initial attack to one node in network A occurs. In **e**, the attacked node is removed, along with its links, and the corresponding interdependent node in network B, with its links, is also removed. In **f**, the actual cascade starts. All the B-links between B-nodes connected interdependently to different A-clusters are removed. In **d**, the same rule applies, but for links in network A: all A-links between A-nodes are deleted if the interdependent nodes do not belong to the same B-cluster. Steps **f** and **g** are repeated, propagating the failures back and forth, until the cascade cannot further evolve. The remaining connected components at the end of the propagation of failures coincide with the mutually connected components introduced in Section 4.5. Figures from [68].

to interlayer ones: in [234], a flexible model including both types of correlations is studied and the maximization of robustness is addressed as a function of the tunable strength of the correlations and the failure propagation dynamics, showing that the theoretical results match well with the experimental results of coupled functional brain modules.

Beyond cascades driven by topological failures, there are other mechanisms via which small perturbations can drive a multilayer network to collapse. In many problems related to social sciences, such as the adoption of fads, the diffusion of norms and innovations, and the changes in the collective attention in a population, models of behavioral contagion are used to model the decisions of the agents. Agents need to decide between two alternative options/actions and the influence of their neighbors is crucial in the final choice. A simple way to encompass this influence is by setting an activation threshold. Initially, all nodes are inactive but a controlled small fraction. An inactive agent changes her state and becomes active only when the fraction of her active neighbors is larger than a threshold. This process might require several activation steps before reaching a frozen state. In [296] was given the range of parameters for which such global cascades, measured as the number of active users when the process stops, emerge in monoplexes, turning out that they are only possible when the network is neither too sparse nor too dense. The generalization of this threshold model in multiplexes was addressed in [65], where a propagation rule is proposed such that an agent activates as far as the fraction of active neighbors in at least one layer exceeds the threshold. Following this rule, multiplexity widens the ranges of parameters in which global cascades can occur, therefore increasing the network's vulnerability. Interestingly, isolated layers in which, owing to their topological properties, global cascades would not be observed, when multiplex-coupled, cooperate in a way that facilitates agent activation and therefore lead to cascades.

Another mechanism that can induce cascades is load redistribution due to overloads. Arguably, the most famous stylized model accounting for this type of dynamics is the Bak-Tang-Wiesenfeld sandpile model [29], a paradigmatic example of self-organized criticality used in a variety of contexts [157]. Each node has an internal, discrete variable – the load. At each unit of time, the load of a node selected uniformly at random increases in one unit. When the load reaches a threshold, assumed to be degree, the node redistributes its load to its neighbors. The neighbors might in turn exceed their capacity and reallo-cate their load to their own neighbors, hence propagating the cascade. Once all nodes have their own load below the threshold, the random addition of load again is restarted. To avoid inundation of the system, a frequently used strategy is to dissipate load at a certain rate when it is reallocated. In [64], the BTW model is used in the context of multilayer power grids, shedding light on the benefits and dangers of the level of interconnectivity between the layers. Sur-prisingly, it is found that there is an optimal value of the interconnectivity for which the chance to observe large cascades is minimum (see Figure 48). This is because the addition of interlayer links between highly isolated power grids

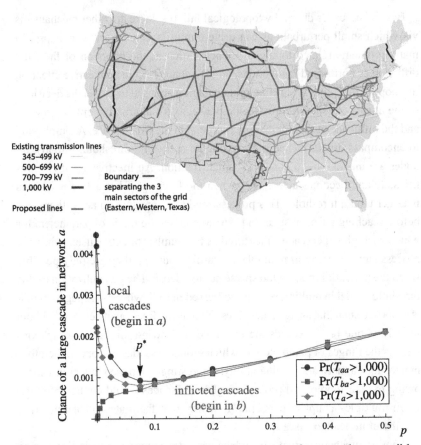

Figure 48 The power grid is a paradigmatic example of a network susceptible to overload cascades of failures due to the energy transported along the links. We see on the top an aggregated representation of the multilayer power grid of the USA, where each layer (link color) corresponds to different voltage ranges. It is depicted the planned interconnections to transport wind power. On the bottom, the Bak-Tang-Wiesenfeld sandpile model is simulated in a duplex of random regular graphs. For a cascade starting in one of the networks (network a) is shown the probability of observing a final cascade size in a, T_{aa}, and in network b, T_{ab}, as a function of the probability of interconnections p between nodes in a and in b. Both T_{aa} and $T_a = 1/2(T_{aa} + T_{ab})$ display a minimum, indicating that an isolated network can reduce the damage caused by a cascade by interconnecting to another network, but only up to a certain level of interconnectedness p^*. Figures readapted from [64].

helps mitigate the cascades by creating a reservoir that absorbs excess load, a sort of alternative dissipative mechanism. This behavior, though, is valid up to a certain point, from which the further addition of interlayer links is no longer beneficial since it creates more loaded systems due to larger thresholds and, therefore, chances of larger cascades, as well as it creates a positive feedback due to the reentrant new paths of load into a layer. Regarding the critical properties of the BTW model, it has been found that multiplexity does not alter the mean-field scaling behavior observed in monoplexes [174].

To close this section, we discuss the relevant, but still quite unexplored, case of nonlocal cascade propagation. One might argue that non-locality can be realized in spatially extended systems by allowing interlayer dependencies between different locations of the space [180]. Yet, under this description, the failures propagate via first neighbors, in the interdependent sense. In fact, all types of cascades discussed earlier evolve locally via first neighbors, something that need not be true in real systems, as happened in the 1996 disturbance of the Western Systems Coordinating Council (WSCC) system [210], in the 2003 blackout in the northeastern USA [200] or in the air traffic disruption due to the eruption of the Icelandic volcano Eyjafjallajökull [117]. A plausible description of this phenomenon is based on the load-capacity model of Motter and Lai [197], where a load, defined as the number of shortest paths crossing a node, and a constant capacity, a factor $1 + \alpha$ larger than the initial load, are assigned to every node. When the load of a node exceeds its capacity, the node fails. An initial perturbation in the form of node removal is applied to the network that globally modifies the loads and allows subsequent failures to occur not necessarily close to the prior failure. If during this process are nodes overload, they also fail, propagating the cascade. This model has been investigated recently

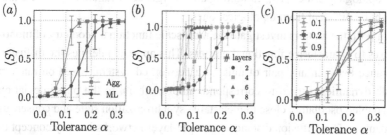

Figure 49 Robustness of multilayer networks exposed to overload failures due to nonlocal load redistribution. We show the behavior of $\langle S \rangle$, the size of the largest connected component of the network at the end of the cascade, as a function of the tolerance parameter α (see text). The faster $\langle S \rangle$ grows, the more robust is the system because there the range of tolerance values for which it disintegrates is smaller. Three mechanisms to increase the robustness are presented. In (a), the comparison between the multilayer network and its aggregated counterpart. In (b), dependence on the number of layers. In (c), $\langle S \rangle$ for different values of the multiplexity parameter (fraction of nodes participating simultaneously in a duplex), with the value indicated in the legend. Figure readapted from [19].

in [19], with the finding that the size of the largest connected component at the end of the cascade suffers an abrupt jump when the tolerance parameter α is increased. Moreover, since the load redistribution depends significantly on the topology of the network, the average path length is identified as a metric that correlates well with robustness. Based on this, the article proposes different strategies to increase the robustness of the network, such as adding new layers or reducing the level of multiplexity (see Figure 49). Unlike the other cascades phenomena introduced in this section, which can be analytically treated with generating functions or multi-type branching processes, nonlocal cascades do not have a solid mathematical machinery to be described, and this certainly represents a challenge for the future.

5 Frontiers

5.1 Kinematic Geometry

The geometric approach [56] to network analysis has garnered growing interest in the past two decades. Here we focus on the geometry of network-driven processes on multilayer networks, a family of kinematic geometries generalizing the diffusion geometry introduced in [90].

The structure of a wide variety of real-world complex systems is modular and hierarchical [144] and the effect of these large-scale properties on the dynamics of such systems has been studied during the past decade. It has been shown that complex systems with such a mesoscale organization [118] are characterized by topological scales [16], exhibiting the emergence of functional clusters that might be different from topological ones.

In [90], the authors investigate the multiscale functional geometry of monoplexes to characterize functional clusters. This approach defines the diffusion distance between any pair of units in a networked system, based on random walk dynamics, shown to correspond, among others, to the phase deviation of coupled oscillators close to metastable synchronization state and consensus dynamics. The diffusion distance for single-layer networks is a key concept to define a kinematic geometry based on network-driven processes and it can be calculated as the L^2−norm of the difference between rows of the propagator $e^{-t\tilde{L}}$:

$$D_t(i,j) = \|\mathbf{p}(t|i) - \mathbf{p}(t|j)\|_2 . \tag{5.1}$$

Exploiting the fact that diffusion geometry is based on random walk dynamics, it is possible to extend it to the realm of multilayer networks. At variance with walks on edge-colored networks presented in Section 4.1, where we can obtain the transition rules for the multigraph as a weighted average of the transition probabilities in each distinct layer (see Equation (4.6)]), on a multilayer network the walk type determines the probability of a random walker jumping across and within layers (see Equation 4.7 and Table 1). Consequently, in the edge-colored case, diffusion distances between nodes can be obtained directly from Equation (5.1), while for multilayers we have to introduce a diffusion distance between state nodes.

Regardless of the random walk type, let us indicate the probability of finding a random walker at a given node and layer at time t by $p_{j\beta}(t)$. Similarly to Equation (5.1), we define the diffusion distance between state nodes (i, α) and (j, β) as

$$D_t^2((i,\alpha),(j,\beta)) = \sum_{k,\gamma}(p_{k\gamma}(t|(i,\alpha)) - p_{k\gamma}(t|(j,\beta)))^2, \tag{5.2}$$

where probabilities are conditional to using the corresponding state nodes as the origins of random walkers at time $t = 0$. One can summarize this supra-distance matrix $\mathbf{D}_t = (D_t((i,\alpha),(j,\beta)))$ into an $N \times N$ matrix by encoding the diffusion distance among the physical nodes, across all the layers. Intuitively, resembling the parallel sum of resistances in electrical circuits leading to the equivalent resistance, the *equivalent diffusion distance* can be written as

$$D_t^{eq}(i,j) = \left(\sum_{\alpha=1}^{L}\frac{1}{D_t((i,\alpha),(j,\alpha))}\right)^{-1}. \tag{5.3}$$

Usually, the functional distance (supra-)matrices are rescaled in $[0, 1]$, normalizing each by its maximum value to allow for comparisons. If, additionally, one is interested in the most persistent patterns, the diffusion distance is averaged over time and called the average diffusion distance. The corresponding supra-distance matrix is indicated by $\bar{\mathbf{D}}_t$.

The diffusion distance between nodes is highly dependent on the type of random walk dynamics, its propagation time, the topology of layers, and the layer-layer correlations. To better understand the effects of dynamics and topological variations, let us consider three classes of two-layer networks, namely:

- Barabasi-Albert layers with preferential attachment of four links;
- Watts-Strogatz layers, with rewiring probability 0.2;
- Girvan-Newman model layers, where the intercommunity connectivity probability is 1, whereas cross-group connections exist with probability 0.05.

Additionally, we consider the five random walks of Table 1. For the three cases, the link overlap across layers – that is, the fraction of links present in both layers among the same pairs of nodes [100] – is fixed to 10%. As shown in Figure 50, the diffusive walk on a scale-free topology leads to a high level of mixed pathways across layers, while in small-world systems, nodes tend to shape more distinct functional clusters.

The functional geometry framework has also been used to analyze empirical systems in [46], where the authors highlighted dissimilarities in the diffusion spaces of the public transportation of London and of the social network of Noordin terrorists by looking at the projections of the spaces and at the results of a Mantel's test [185] applied to supra-distance matrices.

Finally, it is worth mentioning that one can use other walks instead of random ones: this is the case of walks that result from powers of the adjacency matrix and provide the basis for *communicability*, a measure recently used to introduce a geometric framework with applications to single and multiplex networks [115].

5.2 Statistical Physics of Multilayer Systems

5.2.1 Classical Ensembles

To study the properties of an observed multilayer network, comparison with null models is often essential. The maximum entropy approach is one of the standard ways to obtain the required null models, in terms of ensembles of networks exhibiting one (or more) specific property of the observed network, while being maximally random with respect to the other properties. For instance, in the case of single layer networks, one of the most famous null models is the configuration model (CM) [195], which is an ensemble of networks with the same degree sequence as the observed one. The unbiased probability distribution of the members of this ensemble is indicated as $P(G)$ and must maximize the Shannon information entropy

$$S = - \sum_{G \in \mathcal{G}} P(G) \log P(G). \tag{5.4}$$

It is worth mentioning that such a probability distribution does not correspond to a physical process related to the second law of thermodynamics. However, the discussed entropy maximization approach is mimicking the mathematical machinery of statistical physics and, in the case of a fixed degree sequence, it leads to a microcanonical ensemble where all members of the CM have equal probability.

Depending on the type of study, the mentioned constraint of a fixed degree sequence can be relaxed to obtain other ensembles [79]. For instance, by fixing the average degree instead of the full degree sequence, we achieve the grand-canonical ensemble of random graphs. Furthermore, if our knowledge is limited

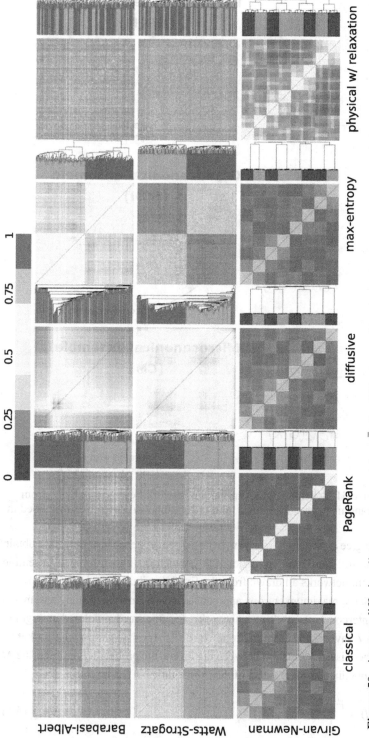

Figure 50 Average diffusion distance supra-matrices $\bar{\mathbf{D}}_t$ for different combinations of multilayer topologies and random walk dynamics (see the text for details). Figure reprinted with permission from [46]. Copyright (2021) by the American Physical Society.

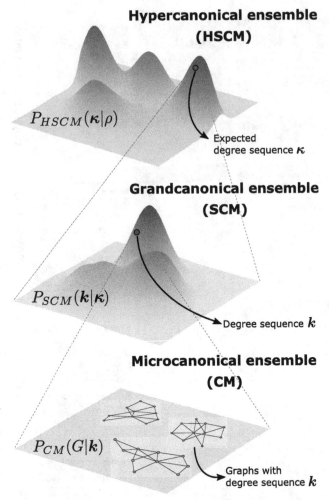

Figure 51 Schematic of sampling the network, represented by G, from hypercanonical, grandcanonical, and microcanonical ensembles discussed in the text. Figure from [290].

to the degree distribution from which the degree sequence is sampled, we obtain the hypersoft configuration model [290], which is a hypercanonical ensemble of random networks all drawn from the fixed distribution (see Figure 51).

Similarly, maximum entropy approaches have been used to study ensembles of noninterconnected multiplex networks. Assume an edge-colored multigraph M with L layers, each forming a network $G^{(\ell)}$ ($\ell = 1, 2, \dots N$). Remarkably, when there is no correlation between the layers, the probability of observing M can be obtained as the product of the probabilities of the layers:

$$P(M) = \prod_{\ell=1}^{L} P(G^{(\ell)}),$$
(5.5)

and the corresponding Shannon entropy becomes the summation of Shannon entropies of layers [49]:

$$S = \sum_{\ell=1}^{L} S^{(\ell)}.$$ (5.6)

By imposing the soft constraints – for example, fixing the average degree instead of degree sequence – and using the Lagrangian multipliers method, we can obtain the canonical multiplex ensemble with probability $P_C(M)$ maximizing the Shannon entropy:

$$P_C(M) = \frac{e^{-\sum_\mu \lambda_\mu F_\mu(M)}}{Z_C} = \prod_{\ell=1}^{L} P_C(G^{(\ell)}),$$ (5.7)

where λ_μ correspond to the Lagrangian multipliers and $F_\mu(M)$ determine the constraints on the network (e.g., average degree). Note that, in the presence of layer-layer correlations, we have $P(M) \neq \prod_{\ell=1}^{L} P(G^{(\ell)})$, leading to other probability distributions extensively studied in [49].

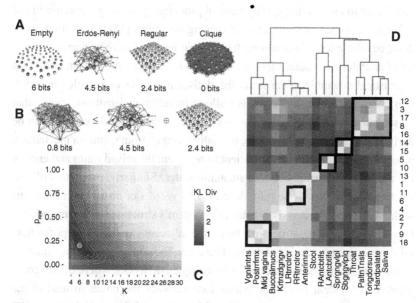

Figure 52 Schematic illustration of the spectral entropy for different classes of networks has been presented. Remarkably, the entropic distances provided by the framework can be used to compare network similarity and characterize the network topology with high accuracy. Figure from [95].

5.2.2 Quantum-Like Ensembles

Complex systems include a wide variety of physical attributes and dynamics. The interactions between their units vary, from the electrochemical signals traveling among neurons in the human brain to the transport of goods between different areas of a urban ecosystem and the spreading of an infectious pathogen between individuals of a society. Regardless of the nature of these interactions, they can be described as information exchange. In fact, complex systems resemble one another in the way they handle information. Therefore, to understand how these systems operate from a physical point of view, investigating their information dynamics is crucial.

It is important to note that the information flow between the units is regulated by the underlying structure, depending on the local neighborhood in certain classes of networks and on long-range communication between distant components in other topologies. At the same time, it is essential to consider the coupling between the structure and the dynamical processes governing the flow of information, such as diffusion, random walks, synchronization, and consensus and, since it is often difficult to understand a system in terms of microscopic features, a framework providing a statistical description of the system might be relevant for applications.

Recently, a statistical field theory of complex information dynamics has been introduced to unify a range of dynamical processes governing the evolution of information on top of static or time-varying structures [132]. This framework describes the interactions among the units in terms of *information streams*, a set of operators that determines the direction of information flow and provides a statistical ensemble to construct the statistical physics of complex information dynamics. In fact, the formalism allows for defining density matrices – that is, statistical average of streams – from which a variety of descriptors can be derived. Of course density matrix formalism comes from quantum mechanics, where vectors were found insufficient to represent the mixed states and encode the pairwise coherence between quantum states. Similarly, properties of networked systems cannot be fully described by vectors or distribution functions without information loss. For instance, a system's structure is often encoded in two-dimensional data structures, being the adjacency matrix, except for trivial symmetries like chains or paths. Therefore, the counterpart of the quantum density matrix has been introduced for complex networks, where off-diagonal elements provide a proxy for interactions between the nodes instead of the coherence between the states. So far, this framework has only been used to analyze classical networks. In the following, we provide a direct application of the theoretical framework to multilayer systems.

Redundancy and Reducibility of Multilayer Systems

As discussed previously, the constituents of complex systems must exchange information efficiently in order to function properly. However, a deep understanding of the structure's role in enhancing or hindering the transport properties – such as navigability [55] – continues to elude us, especially in the case of multilayer systems [53, 97, 103, 129, 136, 161, 231] where an efficient flow might be hindered by the lack of synchronization between different layers [125] or the redundancy of pathways across the layers. There are a number of methods to reduce multiplex networks by structurally merging their most similar layers, based on information-theoretic frameworks [100]. Despite their success, these methods are mostly based on heuristics and have been proved inaccurate under specific circumstances [131].

Enhancing the flow distribution in multilayer systems is challenging since adding links to the layers (e.g., highways, tubes, flights, synapses, etc.) comes with a cost. Interestingly, for multilayer networks with interlinks, changing the weights of interlinks can enhance the diffusion on top of the networks [136]. When acting on the structure is not an option, it has been shown that we can still enhance the transport properties (see Figure 53) using the statistical physics of complex information dynamics by coupling layers dynamically, in a way that a dynamical process cannot distinguish the functionally coupled layers – for example, airlines proposing shared flights to their customers – while evolving. The functional reduction includes coupling the layers with high similarity that

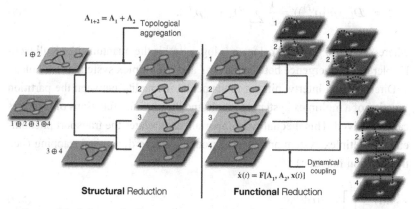

Figure 53 Schematic illustration comparing structural against functional reduction of a multilayer network consisting of $L = 4$ layers. The procedure is similar for both approaches, but the relevant difference is that while structural reducibility alters the topology of the system, function reducibility allows us to functionally couple layers without altering their structure. Figure from [131].

are responsible for the redundant diffusion pathways in the system, in order to identify the (sub)set with maximally diverse layers [131].

Diffusive processes such as random walks have been used extensively to model the information transport within complex structures [188]. Here, we consider random walk dynamics governed by the normalized Laplacian $\hat{\bar{\mathbf{L}}} = \langle \hat{\mathbf{L}}^{(\ell)} \rangle$ given by $\hat{\bar{\mathbf{L}}} = \hat{\mathbf{I}} - \langle \hat{\mathbf{T}}^{(\ell)} \rangle$ playing the role of the quasi-Hamiltonian (see Equation [4.6]). The density matrix can be obtained for random walks on multiplex networks as $\hat{\rho}(t) = \frac{e^{-t\hat{\bar{\mathbf{L}}}}}{Z(t)}$.

Interestingly, it has been shown that the partition function, which is encoding the amount of the trapped field, is proportional to the average return probability of random walk dynamics: $Z(t) = N\mathcal{R}(t)$, where $\mathcal{R}(t) = N^{-1} \sum_{i=1}^{N} e^{-t\lambda_i}$ is the average return probability and λ_i is the i-th eigenvalue of $\hat{\bar{\mathbf{L}}}$. Intuitively, the average return probability is high when the structural symmetries and abundance of redundant diffusion pathways slow down the information propagation between the units. Thus, it is expected that breaking the structural regularities by adding long-range interactions [297] or increasing the diversity of diffusion pathways across layers can lead to faster information flow and and lower $Z(t)$.

Multiplexity of interactions among units generates nontrivial dynamical correlations between layers that have no counterpart when layers are considered in isolation. This important difference can be characterized in terms of average entropy distance between the multiplex and its layers. This measure, named intertwining, is defined by

$$I = \langle \mathcal{D}_{KL}(\hat{\rho}||\hat{\rho}^{(\ell)}) \rangle = \frac{1}{L} \sum_{\ell=1}^{L} \mathcal{D}_{KL}(\hat{\rho}||\hat{\rho}^{(\ell)}), \tag{5.8}$$

where $\mathcal{D}_{KL}(\hat{\rho}||\hat{\rho}^{(\ell)}) = Tr[\hat{\rho}(\log_2 \hat{\rho} - \log_2 \hat{\rho}^{(\ell)})]$ is the quantum-like Kullback-Leibler (KL) divergence between layer ℓ and a multiplex system as a whole.

Directly from intertwining, a fundamental inequality between the partition function of a multiplex system as whole and the partition functions of its layers can be derived. This inequality is important as it relates the transport phenomena of multiplex system and layers through average dynamical trapping (i.e., the partition function):

$$Z(t) \leq \prod_{\ell=1}^{L} Z^{(\ell)}(t)^{1/L}, \tag{5.9}$$

where equality holds if and only if all the layers are the same. Using dynamical trapping as a measure of transport, Equation (5.9) shows that a multiplex network has better transport properties than the geometric average of layers, adding an advantage to multilayer structures.

Furthermore, being reminiscent of statistical physics of particles, the equality defines the noninteracting scenario where layer-layer correlations do not alter the underlying dynamics: the entropy $S^{(\ell)}(t)$ of each layer is calculated separately and the overall entropy is given by their average $S_{nint}(t) = \langle S^{(\ell)}(t) \rangle$. Conversely, any topological alteration of the noninteracting scenario introduces a dynamical correlation between layers, requiring the exploration of layers to gather more information about the system: in this case, the network consists of interacting layers where the diffusion dynamics on the whole multiplex network is considered to measure the entropy $S_{int}(t)$. To obtain another form of Equation (5.8), a mean-field approximation of the Von Neumann entropy is given by

$$S^{MF}(t) = \frac{1}{\log 2}\left(t\frac{Z(t)-1}{Z(t)} + \log Z(t)\right),\tag{5.10}$$

and can be used to prove that layer-layer interactions lower the system's entropy $(S_{int}(t) \le S_{nint}(t))$ and partition function $Z(t)$. Normalizing the intertwining by its upper bound, for values of time t sufficiently large and in absence of isolated state nodes, Equation (5.8) reduces to the relative intertwining:

$$\mathcal{I}^*(t) = 1 - \frac{S_{int}(t)}{S_{nint}(t)},\tag{5.11}$$

which is bounded between 0 (i.e., the layers are redundant) and 1 (i.e., the layers are diverse and the system is irreducible).

We can show how intertwining is proportional to the functional diversity of layers. The Laplacian matrix of the multiplex network is given by $\hat{\bar{L}} = \langle \hat{L}^{(\ell)} \rangle$. Therefore, the Laplacian matrix of each layer $\hat{L}^{(\ell)}$ can be written as a perturbation of multiplex Laplacian $\hat{L}^{(\ell)} = \hat{\bar{L}} + \Delta\hat{L}^{(\ell)}$, reflected in its eigenvalues as $\lambda_i^{(\ell)} = \bar{\lambda}_i + \Delta\lambda_i^{(\ell)}$ ($i = 0, 1, \ldots N$). It is straightforward to show that $\frac{1}{N}\sum_{i=1}^{N}\Delta\lambda_i^{(\ell)} = \overline{\Delta\lambda^{(\ell)}} = 0$ and that $\overline{\Delta\lambda^{(\ell)2}} \ge 0$, the latter quantifying the influence of the perturbation to each layer. The average of the variance across all layers $\langle(\Delta\lambda^{(\ell)})^2\rangle$ provides a measure of the overall spectral diversity of layers, which, interestingly, is demonstrated to be proportional to the relative intertwining:

$$\mathcal{I}^*(t) \approx \frac{t^2}{2}\langle(\Delta\lambda^{(\ell)})^2\rangle.\tag{5.12}$$

This proves the sensitivity of intertwining to the spectral diversity of layers. Furthermore, the partition functions of layers can be written in terms of perturbations such as $Z^{(\ell)}(t) = Z(t) + \Delta Z^{(\ell)}(t)$, leading to:

$$\mathcal{I}^*(t) \approx \frac{\overline{\Delta Z^{(\ell)}(t)}}{Z(t)-1}; \quad \overline{\Delta Z^{(\ell)}(t)} = \frac{1}{L}\sum_{\ell=1}^{L}\Delta Z^{(\ell)}(t).\tag{5.13}$$

Equations (5.12) and (5.13) provide a fundamental result: they show that by minimizing the partition function of the system, we maximize the relative intertwining while favoring the maximum functional diversity of layers. Additionally, it has been shown that an inverse proportionality holds between the diffusion time $(1/\lambda_2)$ and intertwining, as further evidence for the role of intertwining in characterizing the transport properties.

These theoretical findings have been applied to a broad range of synthetic and empirical systems, providing a transparent framework for coupling the most similar layers of multiplex networks in order to improve their transport properties including dynamical trapping, diffusion time, and navigability [131].

6 Conclusions

Network science is one of the greatest achievements of the twenty-first century, paving the way toward a mathematical approach for the analysis of disparate complex systems and allowing us to find regularities in apparently disordered connectivity patterns. The past decade has seen the flourishing of analytical techniques and models exploiting or characterizing the inherent multidimensionality of empirical systems, from multiplexity – that is, the existence of distinct types of relationships among the same set of actors or units – to interdependency – that is, the existence of structural or functional connections among sets of actors or units of a different nature. Such multiple dimensions are nowadays easily encoded into layers of information.

Most of the advances in this direction are described in some detail or referred to in this work. Starting from the mathematical representation of multilayer networks (Section 2), we have introduced structural descriptors for units, layers, and the whole system (Section 3), providing an overview of the micro- and mesoscale organization of such systems. We have discussed the rich spectrum of phenomena, with no classical counterparts, related to dynamical processes on the top of the networks and their intertwining (Section 4), guiding the reader toward two promising research areas for the future, namely network geometry and information dynamics (Section 5), although many other exciting subfields are emerging – for example, higher-order modeling and analysis [38, 170].

At this point, the reader should be sufficiently familiar with multilayer network science and, to conclude, we would like to make a quick journey through the most recent applications of its paradigm, moving across different spatial scales and ranging from cells to societies.

The first stop of this journey is exactly a cell, which can vary in diameter between 10^{-6} m and 10^{-4} m (note that a DNA double helix is about 10^{-8} m wide). The cell can be seen as a multilayer system consisting of three interdependent layers (see Figure 54). Here, applications are mostly related to

The cell as a multilayer network

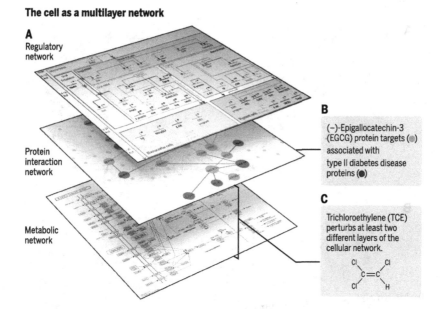

A Regulatory network

B (–)-Epigallocatechin-3 (EGCG) protein targets (●) associated with type II diabetes disease proteins (●)

Protein interaction network

C Trichloroethylene (TCE) perturbs at least two different layers of the cellular network.

Metabolic network

Figure 54 Multilayer representation of a cell. Layers: (i) regulatory interactions involving RNA and protein expression, (ii) protein-protein interactions involved in signaling and responsible cell function, (iii) metabolic interactions with reactions and pathways crucial for cell function. Figure from [287].

the emerging field of systems biology and network medicine, promising to use biomolecular interactions to develop a deeper knowledge of biology across scales with the ultimate goal to better understand diseases, to prevent them, and to treat them with medical drugs that reduce side effects.

The second stop requires a big jump of about three orders of magnitude, to explore one of the most famous multicellular organisms with a small-scale neural system: the *Caenorhabditis elegans*, a nematode of about 10^{-3} m. The nervous system of this small worm consists of synaptic and neuropeptide interactions that can be seen as interconnected layers. Here, the multilayer perspective provides the opportunity to better understand how the integration between hard-wired synaptic or junctional circuits and extrasynaptic signals can modulate the large-scale behavior of the worm [44].

At a larger scale, around 10^{-1} m, another emblematic neural system, namely the human brain, is currently being characterized in terms of how its structural and functional connectivity evolves over time – for example, while performing a specific task, or across groups, unraveling the existence of modular and hierarchical structures that would remain hidden under the lens of less sophisticated models and analytical techniques [33]. In parallel, the functional

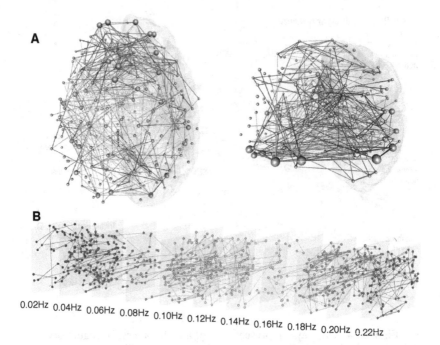

0.02Hz 0.04Hz 0.06Hz 0.08Hz 0.10Hz 0.12Hz 0.14Hz 0.16Hz 0.18Hz 0.20Hz 0.22Hz

Figure 55 Multilayer representation of a human brain. Both the 3D and layered visualization encode the functional brain of a schizophrenic subject (11 nonoverlapping frequency-band layers between 0.01 and 0.23 Hz). Figure from [91], readapted from [104].

connectivity of the human brain can be stratified by frequency bands (see Figure 55) where specific correlations or causal relationships between regions of interests appear [91]. Multilayer analysis can be used to better characterize the relative importance of all brain regions [298] and to enhance the accuracy in discriminating between healthy and unhealthy patients starting from imaging information [104].

At larger scales, on the order of hundreds of meters (10^2 m), cooperative systems like that of dolphins can be characterized with respect to distinct types of interactions, allowing us to get unprecedented insights about the underlying social organization and group dynamics. At similar spatial scales, another emblematic example, accounting for the socio-spatial interdependence typical of many other networks, concerns the organization of ecological systems (Figure 56): for instance, it has been recently shown that the importance of dispersers for an ecosystem is better captured by multilayer measures of importance rather than by standard metrics [277].

At the human scale, relatively small-scale social systems (10^4 m) have been studied to better understand the impact of external shocks on social structure and dynamics. For instance, it has been shown that multilayer modeling can

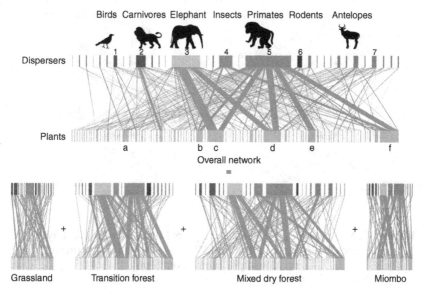

Figure 56 Aggregate (top) and multilayer (bottom) representations of a real seed-dispersal network. Reproduced from [277] under Creative Commons Attribution 4.0 International License.

disentangle the different social ties that conform to the proximity interaction networks of extant hunter-gatherer societies, identifying social relationships with a key role in the spread and accumulation of culture [191]. In another study, the analysis of mixed economies in three villages of Alaska unraveled that factors related to climate change, such as global warming, have a non-negligible effect on household structure, but the most important factors for vulnerability are indeed due to social shift rather than resource depletion [28].

Let us jump by three more orders of magnitude and discuss systems at the planetary scale: the ones based on information exchange thanks to large-scale communication infrastructures, such as the Internet. At this scale, the activity of a complex system might be very frenetic; think about an online social media platform where millions or billions of users worldwide continuously produce content to be shared, which quickly travels and bounces from one country to another. For instance, it is interesting to study how rumors and memes spread on these systems, as recently proposed in [86] to capture the behavior of users who post information from one social media platform to another and to provide a plausible explanation for the heavy-tailed distribution of meme popularity that is usually observed in empirical data.

Recent applications also include the analysis of trade networks and their nested and modular structure [1, 9, 278]. Additionally, as anticipated, multilayer models of financial networks have been proposed very recently to better explain how the financial distress of one country can ignite a global

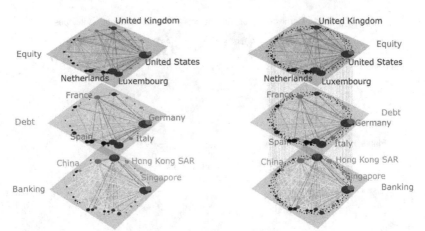

Figure 57 Multilayer representation of a global financial network. Layers are asset types, nodes are countries, and links are cross-country financial relationships. Figure from [106].

financial crisis, like the one in 2008. By considering distinct asset types as layers where countries (nodes) exchange financial assets (links), [106] have highlighted the importance of both intralayer and interlayer connectivity contributions to the propagation of contagions (Figure 57).

The long journey summarized in the last few pages of this work allowed us to stress the importance of multilayer modeling across a broad spectrum of disciplines, including cell biology, neuroscience, ecology, and social sciences, spanning 10 orders of magnitude from a spatial scale of about 10^{-6} m to one of about 10^4 m. The theoretical and computational framework presented in this work provides researchers and practitioners with a versatile and unified tool kit to shed light and gain new insights on the complexity of natural and artificial systems.

Appendix A
Master Stability Function (MSF) Formalism

To test the stability of the synchronized state $\mathbf{s}$ we study how the perturbation error $\delta\mathbf{x}_i(t) = \mathbf{x}_i(t) - \mathbf{s}(t)$ evolves. By assuming small perturbations, we can write the variational equation for $\delta\mathbf{x}_i$:

$$\delta\mathbf{x}_i = J\mathbf{F}(\mathbf{s})\delta\mathbf{x}_i - \sigma J\mathbf{H}(\mathbf{s}) \sum_{j=1}^{N} L_{ij}\delta\mathbf{x}_j \tag{A.1}$$

where $J\mathbf{F}$ and $J\mathbf{H}$ are the Jacobian of $\mathbf{F}$ and $\mathbf{H}$, respectively.

To find the solution of Equation (A.1) we have to project $\delta\mathbf{x}$ into the eigenspace formed by the eigenvectors of the Laplacian matrix $\mathbf{L}$, obtaining a decomposition of the time evolution of perturbation error into N decoupled eigenmodes:

$$\dot{\boldsymbol{\xi}}_i = [J\mathbf{F}(\mathbf{s}) - \sigma\lambda_i J\mathbf{H}(\mathbf{s})]\boldsymbol{\xi}_i, \quad i = 1,\ldots,N, \tag{A.2}$$

where $\boldsymbol{\xi}_i$ is the eigenmode associated with the eigenvalue λ_i of $\mathbf{L}$. By indicating as Λ_{max} the maximum Lyapunov exponent associated with the system of equations (A.2), we can write the time evolution of $\boldsymbol{\xi}$ as $|\boldsymbol{\xi}| \sim e^{\Lambda_{max}t}$ and, finally, we found that a necessary condition for the stability of a synchronized state is that $\Lambda_{max} < 0$. The expression of Λ_{max} as a function of a generic parameter $\alpha_i = \sigma\lambda_i$ is named the master stability function (MSF), and it usually assumes negative values in an interval $\alpha \in (\alpha_1, \alpha_2)$. It means that, for a fixed coupling strength σ, a network can reach and maintain complete synchronization only if its structure, defined by its Laplacian matrix, is such that:

$$\alpha_1 < \sigma\lambda_2 \leq \sigma\lambda_{\leq} \ldots \leq \sigma\lambda_N < \alpha_2. \tag{A.3}$$

or, equivalently,

$$R \equiv \frac{\lambda_N}{\lambda_2} < \frac{\alpha_2}{\alpha_1}, \tag{A.4}$$

used, for instance, in [258] to unravel the existence of an optimal value for the synchronizability of a multilayer system.

Appendix B
Kuramoto Model on Networks

The first approach to model collective synchronization considers a population of coupled limit-cycle oscillators whose natural frequencies are drawn from some prescribed distribution and exert a phase-dependent influence on each other. In formulae, we can write the Kuramoto model (KM) [17, 270] as a system of N oscillators whose instantaneous phases θ_i are described by the equation:

$$\dot{\theta}_i = \omega_i + \frac{\sigma}{N} \sum_{j=1}^{N} \sin(\theta_j - \theta_i), \tag{B.1}$$

where σ is the coupling constant and ω_i is the natural frequency of oscillator i, chosen from an unimodal distribution $g(\omega)$.

We can use a order parameter that describes the transition: it is defined as a macroscopic measure that quantifies the collective rhythm produced by the whole population:

$$r(t)e^{i\Phi(t)} = \frac{1}{N} \sum_{j=1}^{N} \sin(\theta_j), \tag{B.2}$$

where $\Phi(t)$ is the average phase and $0 \leq r(t) \leq 1$ measures the phase coherence, where the two extreme values correspond to phase-locked ($r = 1$) or incoherent oscillators. Equation (B.1) is then rewritten in terms of the order parameter, and the equation for the instantaneous phase reduces to:

$$\dot{\theta}_i = \omega_i + \sigma r \sin(\theta_i), \tag{B.3}$$

which has two types of long-term behaviors. Oscillators for which $|\omega_i| \leq \sigma r$ are phase-locked and form a mutually synchronized cluster. Oscillators with frequencies in the tails of $g(\omega)$ distribution, where $|\omega_i| > \sigma r$ holds, are drifting with respect to the synchronized cluster.

The Kuramoto model is then generalized on networks by including in Equation (B.1) information about network connectivity:

$$\dot{\theta}_i = \omega_i + \sum_{j=1}^{N} \sigma_{ij} A_{ij} \sin(\theta_j - \theta_i). \tag{B.4}$$

Appendix C

Transitions in Multilayer Systems

Table C.1 Phase transitions (C: continuous, D: discontinuous, H: hybrid) in multilayer networks due to the algebraic, topological, and simple dynamics described in Section 4.1.

Type of phase transition	Mode	Reference
Enhanced diffusion	C	[136]
Emergence of multiplexity	D	[231]

Table C.2 Phase transitions in multilayer networks due to the simple and couple dynamics described in Sections 4.2, 4.3, and 4.4.

Type of dynamics	Mode	Reference
Synchronization	C	[128]
Explosive synchronization with local-order parameter	D	[305]
Synchronization of oscillators with random walks	C & D	[208]
Interacting diseases	C	[249]
Epidemic onset control by raising of awareness	C	[141]
Explosive pandemics with no early warning	D	[88]
Cooperative behavior in coupled competitive games – social influence	C	[10]
Cooperative behavior in prisoner's dilemma game	C	[139]
Cooperative behavior in public good game	C	[37]

Table C.3 Type of transitions for different scenarios in percolation and cascades propagation

Type of network	Mode	Reference
Multilayer network	C	[177]
Multiplex	D	[263]
Partial multiplex	C & D	[263]

Table C.3

Multiplex with overlap	C, D & H	[40, 75, 150]
Multiplex with spatially embedded networks	C & D	[32, 143, 262]
Bond percolation on multiplex	C	[145]
Multiplex & local cascades	D	[68]
Multiplex & nonlocal cascades	D	[19]

References

[1] L. G. A. Alves, G. Mangioni, F. A. Rodrigues, P. Panzarasa, and Y. Moreno. Unfolding the complexity of the global value chain: Strength and entropy in the single-layer, multiplex, and multi-layer international trade networks. *Entropy*, 20 (12): 909, 2018.

[2] J. A. Acebrón, L. L. Bonilla, C. J. P. Vicente, F. Ritort, and R. Spigler. The Kuramoto model: A simple paradigm for synchronization phenomena. *Reviews of Modern Physics*, 77 (1): 137, 2005.

[3] S. Achard and E. Bullmore. Efficiency and cost of economical brain functional networks. *PLoS Computational Biology*, 3 (2): e17, 2007.

[4] L. A. Adamic and E. Adar. Friends and neighbors on the Web. *Social Networks*, 25 (3): 211–230, 2003.

[5] E. M. Airoldi, D. M. Blei, S. E. Fienberg, and E. P. Xing. Mixed membership stochastic blockmodels. *Journal of Machine Learning Research*, 9: 1981–2014, 2008.

[6] M. Akbarzadeh and E. Estrada. Communicability geometry captures traffic flows in cities. *Nature Human Behaviour*, 2 (9): 645–652, 2018.

[7] R. Albert, H. Jeong, and A.-L. Barabási. Error and attack tolerance of complex networks. *Nature*, 406 (6794): 378, 2000.

[8] A. Aleta, M. Tuninetti, D. Paolotti, Y. Moreno, and M. Starnini. Link prediction in multiplex networks via triadic closure. *Physical Review Research*, 2 (4): 042029, 2020.

[9] L. G. A. Alves, G. Mangioni, I. Cingolani, F. A. Rodrigues, P. Panzarasa, and Y. Moreno. The nested structural organization of the worldwide trade multi-layer network. *Scientific Reports*, 9 (1): 1–14, 2019.

[10] R. Amato, A. Díaz-Guilera, and K.-K. Kleineberg. Interplay between social influence and competitive strategical games in multiplex networks. *Scientific Reports*, 7 (1): 1–8, 2017.

[11] R. Amato, N. E. Kouvaris, M. San Miguel, and A. Díaz-Guilera. Opinion competition dynamics on multiplex networks. *New Journal of Physics*, 19 (12): 123019, 2017.

[12] A. Amelio, G. Mangioni, and A. Tagarelli. Modularity in multilayer networks using redundancy-based resolution and projection-based inter-layer coupling. *IEEE Transactions on Network Science and Engineering*, 7(3):1198–1214, 1 July–Sept. 2020. https://doi.org/10.1109/TNSE.2019.2913325.

[13] A. Anandkumar, R. Ge, D. Hsu, and S. M. Kakade. A tensor approach to learning mixed membership community models. *Journal of Machine Learning Research*, 15 (1): 2239–2312, 2014.

[14] P. W. Anderson. More is different. *Science*, 177 (4047): 393–396, 1972.

[15] C. G. Antonopoulos and Y. Shang. Opinion formation in multiplex networks with general initial distributions. *Scientific Reports*, 8 (1): 2852, 2018.

[16] A. Arenas, A. Díaz-Guilera, and C. J. Pérez-Vicente. Synchronization reveals topological scales in complex networks. *Physical Review Letters*, 96: 114102, 2006.

[17] A. Arenas, A. Diaz-Guilera, J. Kurths, Y. Moreno, and C. Zhou. Synchronization in complex networks. *Physics Reports*, 469(3):93–153, 2008. https://doi.org/10.1016/j.physrep.2008.09.002

[18] A. Arenas, A. Fernandez, and S. Gomez. Analysis of the structure of complex networks at different resolution levels. *New Journal of Physics*, 10 (5): 053039, 2008.

[19] O. Artime and M. De Domenico. Abrupt transition due to non-local cascade propagation in multiplex systems. *New Journal of Physics*, 22 (9): 093035, 2020.

[20] O. Artime and M. De Domenico. Percolation on feature-enriched interconnected systems. *Nature Communications*, 12 (1): 1–12, 2021.

[21] O. Artime, J. Fernández-Gracia, J. J. Ramasco, and M. San Miguel. Joint effect of ageing and multilayer structure prevents ordering in the voter model. *Scientific Reports*, 7 (1): 7166, 2017.

[22] O. Artime, V. d'Andrea, R. Gallotti, P. L. Sacco, and M. De Domenico. Effectiveness of dismantling strategies on moderated vs. unmoderated online social platforms. *Scientific Reports*, 10 (1): 1–11, 2020.

[23] U. Aslak, M. Rosvall, and S. Lehmann. Constrained information flows in temporal networks reveal intermittent communities. *Physical Review E*, 97 (6): 062312, 2018.

[24] M. Asllani, D. M. Busiello, T. Carletti, D. Fanelli, and G. Planchon. Turing patterns in multiplex networks. *Physical Review E*, 90 (4): 042814, 2014.

[25] M. Asllani, D. M. Busiello, T. Carletti, D. Fanelli, and G. Planchon. Turing instabilities on Cartesian product networks. *Scientific Reports*, 5 (1): 1–10, 2015.

[26] N. Azimi-Tafreshi. Cooperative epidemics on multiplex networks. *Physical Review E*, 93 (4): 042303, 2016.

[27] N. Azimi-Tafreshi, J. Gómez-Gardenes, and S. N. Dorogovtsev. k–Core percolation on multiplex networks. *Physical Review E*, 90 (3): 032816, 2014.

[28] J. A. Baggio, S. B. BurnSilver, A. Arenas, J. S. Magdanz, G. P. Kofinas, and M. De Domenico. Multiplex social ecological network analysis reveals how social changes affect community robustness more than resource depletion. *Proceedings of the National Academy of Sciences*, 113 (48): 13708–13713, 2016.

[29] P. Bak, C. Tang, and K. Wiesenfeld. Self-organized criticality. *Physical Review A*, 38 (1): 364, 1988.

[30] A.-L. Barabási and M. Pósfai. *Network science*. Cambridge University Press, 2016.

[31] A. Barrat, M. Barthelemy, and A. Vespignani. *Dynamical processes on complex networks*. Cambridge University Press, 2008.

[32] A. Bashan, Y. Berezin, S. V. Buldyrev, and S. Havlin. The extreme vulnerability of interdependent spatially embedded networks. *Nature Physics*, 9 (10): 667–672, 2013.

[33] D. S. Bassett and O. Sporns. Network neuroscience. *Nature Neuroscience*, 20 (3): 353, 2017.

[34] F. Battiston, V. Nicosia, and V. Latora. Structural measures for multiplex networks. *Physical Review E*, 89 (3): 032804, 2014.

[35] F. Battiston, V. Nicosia, and V. Latora. Efficient exploration of multiplex networks. *New Journal of Physics*, 18 (4): 043035, 2016.

[36] F. Battiston, V. Nicosia, and V. Latora. The new challenges of multiplex networks: measures and models. *European Physical Journal Special Topics*, 226 (3): 401–416, 2017.

[37] F. Battiston, M. Perc, and V. Latora. Determinants of public cooperation in multiplex networks. *New Journal of Physics*, 19 (7): 073017, 2017.

[38] F. Battiston, G. Cencetti, I. Iacopini, et al. Networks beyond pairwise interactions: Structure and dynamics. *Physics Reports*, 874:1–92, 2020.

[39] G. J. Baxter, S. N. Dorogovtsev, A. V. Goltsev, and J. F. F. Mendes. Avalanche collapse of interdependent networks. *Physical Review Letters*, 109 (24): 248701, 2012.

[40] G. J. Baxter, G. Bianconi, R. A. da Costa, S. N. Dorogovtsev, and J. F. F. Mendes. Correlated edge overlaps in multiplex networks. *Physical Review E*, 94 (1): 012303, 2016.

[41] M. Bazzi, M. A. Porter, S. Williams, M. McDonald, D. J. Fenn, and S. D. Howison. Community detection in temporal multilayer networks, with an application to correlation networks. *Multiscale Modeling & Simulation*, 14 (1): 1–41, 2016.

[42] M. Bazzi, L. G. S. Jeub, A. Arenas, S. D. Howison, and M. A. Porter. A framework for the construction of generative models for mesoscale structure in multilayer networks. *Physical Review Research*, 2 (2): 023100, 2020.

[43] B. Beisner, N. Braun, M. Pósfai, J. Vandeleest, R. D'Souza, and B. McCowan. A multiplex centrality metric for complex social networks: Sex, social status, and family structure predict multiplex centrality in rhesus macaques. *PeerJ*, 8: e8712, 2020.

[44] B. Bentley, R. Branicky, C. L. Barnes, et al. The multilayer connectome of *Caenorhabditis elegans. PLoS Computational Biology*, 12 (12): e1005283, 2016.

[45] Y. Berezin, A. Bashan, and S. Havlin. Comment on "Percolation transitions are not always sharpened by making networks interdependent." *Physical Review Letters*, 111 (18): 189601, 2013.

[46] G. Bertagnolli and M. De Domenico. Diffusion geometry of multiplex and interdependent systems. *Physical Review E*, 103: 042301, 2021.

[47] G. Bertagnolli, R. Gallotti, and M. De Domenico. Quantifying efficient information exchange in real network flows. *Communications Physics*, 4 (1): 1–10, 2021.

[48] J. Biamonte, M. Faccin, and M. De Domenico. Complex networks from classical to quantum. *Communications Physics*, 2 (1): 1–10, 2019.

[49] G. Bianconi. Statistical mechanics of multiplex networks: Entropy and overlap. *Physical Review E*, 87: 062806, 2013.

[50] G. Bianconi. Epidemic spreading and bond percolation on multilayer networks. *Journal of Statistical Mechanics*, 2017 (3): 034001, 2017.

[51] G. Bianconi. *Multilayer networks: Structure and function*. Oxford University Press, 2018.

[52] G. Bianconi and F. Radicchi. Percolation in real multiplex networks. *Physical Review E*, 94 (6): 060301, 2016.

[53] S. Boccaletti, G. Bianconi, R. Criado, et al. The structure and dynamics of multilayer networks. *Physics Reports*, 544 (1): 1–122, 2014.

[54] S. Boccaletti, A. N. Pisarchik, C. I. del Genio, and A. Amann. *Synchronization*. Cambridge University Press, 2018.

[55] M. Boguñá, D. Krioukov, and K. C. Claffy. Navigability of complex networks. *Nature Physics*, 5 (1): 74, 2009.

[56] M. Boguñá, I. Bonamassa, M. De Domenico, S. Havlin, D. Krioukov, and M. Á. Serrano. Network geometry. *Nature Reviews Physics*, 3:114–135, 2021.

[57] P. Bonacich. Power and centrality: A family of measures. *American Journal of Sociology*, 92 (5): 1170–1182, 1987.

[58] S. P. Borgatti and M. G. Everett. A graph-theoretic perspective on centrality. *Social Networks*, 28 (4): 466–484, 2006.

[59] P. Bosetti, P. Poletti, M. Stella, B. Lepri, S. Merler, and M. De Domenico. Heterogeneity in social and epidemiological factors determines the risk of measles outbreaks. *PNAS*, 117:30118, 2020.

[60] L. Bottcher and M. A. Porter. Classical and quantum random-walk centrality measures in multilayer networks. *arxiv preprint arXiv:2012.07157*, 2020.

[61] A. Brechtel, P. Gramlich, D. Ritterskamp, B. Drossel, and T. Gross. Master stability functions reveal diffusion-driven pattern formation in networks. *Physical Review E*, 97 (3), 2018.

[62] S. Brin and L. Page. The anatomy of a large-scale hypertextual web search engine. *Computer Networks and ISDN Systems*, 30 (1–7): 107–117, 1998.

[63] P. Bródka, A. Chmiel, M. Magnani, and G. Ragozini. Quantifying layer similarity in multiplex networks: A systematic study. *Royal Society Open Science*, 5 (8): 171747, 2018.

[64] C. D. Brummitt, R. M. D'Souza, and E. A. Leicht. Suppressing cascades of load in interdependent networks. *PNAS*, 109 (12): E680–E689, 2012.

[65] C. D. Brummitt, K.-M. Lee, and K.-I. Goh. Multiplexity-facilitated cascades in networks. *Physical Review E*, 85 (4): 045102, 2012.

[66] F. Buccafurri, G. Lax, S. Nicolazzo, A. Nocera, and D. Ursino. Measuring betweenness centrality in social internetworking scenarios. In Y. T. Demey and H. Panetto (eds.), *On the move to meaningful internet systems: OTM 2013 Workshops. OTM 2013. Lecture Notes in Computer Science, vol. 8186*. Springer, 2013. https://doi.org/10.1007/978-3-642-41033-8_84

[67] V. Buendía, P. Villegas, R. Burioni, and M. A. Muñoz. The broad edge of synchronisation: Griffiths effects and collective phenomena in brain networks. *arXiv preprint arXiv:2109.11783*, 2021.

[68] S. V. Buldyrev, R. Parshani, G. Paul, H. E. Stanley, and S. Havlin. Catastrophic cascade of failures in interdependent networks. *Nature*, 464 (7291): 1025–1028, 2010.

[69] C. Buono, L. G. Alvarez-Zuzek, P. A. Macri, and L. A. Braunstein. Epidemics in partially overlapped multiplex networks. *PloS One*, 9 (3): e92200, 2014.

[70] Z. Burda, J. Duda, J.-M. Luck, and B. Waclaw. Localization of the maximal entropy random walk. *Physical Review Letters*, 102 (16): 160602, 2009.

[71] D. M. Busiello, T. Carletti, and D. Fanelli. Homogeneous-per-layer patterns in multiplex networks. *Europhysics Letters*, 121 (4): 48006, 2018.

[72] V. Carchiolo, A. Longheu, M. Malgeri, and G. Mangioni. Communities unfolding in multislice networks. In *Complex Networks*, pages 187–195. Springer, 2011.

[73] A. Cardillo, J. Gómez-Gardeñes, M. Zanin, et al. Emergence of network features from multiplexity. *Scientific Reports*, 3 (1), 2013.

[74] D. Cellai, E. López, J. Zhou, J. P. Gleeson, and G. Bianconi. Percolation in multiplex networks with overlap. *Physical Review E*, 88 (5): 052811, 2013.

[75] D. Cellai, S. N. Dorogovtsev, and G. Bianconi. Message passing theory for percolation models on multiplex networks with link overlap. *Physical Review E*, 94 (3): 032301, 2016.

[76] D. Centola. The social origins of networks and diffusion. *American Journal of Sociology*, 120 (5): 1295–1338, 2015.

[77] P. S. Chodrow, Z. Al-Awwad, S. Jiang, and M. C. González. Demand and congestion in multiplex transportation networks. *PloS One*, 11 (9): e0161738, 2016.

[78] F. R. K. Chung. *Spectral graph theory*. 2nd edition. American Mathematical Society, 1997.

[79] G. Cimini, T. Squartini, F. Saracco, et al. The statistical physics of real-world networks. *Nature Reviews Physics*, 1 (1): 58–71, 2019.

[80] R. Cohen, K. Erez, D. Ben-Avraham, and S. Havlin. Breakdown of the Internet under intentional attack. *Physical Review Letters*, 86 (16): 3682, 2001.

[81] E. Cozzo, R. A. Baños, S. Meloni, and Y. Moreno. Contact-based social contagion in multiplex networks. *Physical Review E*, 88 (5): 050801, 2013.

[82] E. Cozzo, M. Kivelä, M. De Domenico, et al. Structure of triadic relations in multiplex networks. *New Journal of Physics*, 17 (7): 073029, 2015.

[83] E. Cozzo, G. F. De Arruda, F. A. Rodrigues, and Y. Moreno. *Multiplex networks: Basic formalism and structural properties*. Springer, 2018.

[84] R. Criado, J. Flores, A. García del Amo, J. Gómez-Gardeñes, and M. Romance. A mathematical model for networks with structures in the mesoscale. *International Journal of Computer Mathematics*, 89 (3): 291–309, 2012.

[85] A. Czaplicka, R. Toral, and M. San Miguel. Competition of simple and complex adoption on interdependent networks. *Physical Review E*, 94 (6): 062301, 2016.

[86] J. D. O'Brien, I. K. Dassios, and J. P. Gleeson. Spreading of memes on multiplex networks. *New Journal of Physics*, 21 (2): 025001, 2019.

[87] M. M. Danziger, L. M. Shekhtman, A. Bashan, Y. Berezin, and S. Havlin. Vulnerability of interdependent networks and networks of networks. In *Interconnected Networks*, pages 79–99. Springer, 2016.

[88] M. M. Danziger, I. Bonamassa, S. Boccaletti, and S. Havlin. Dynamic interdependence and competition in multilayer networks. *Nature Physics*, 15 (2): 178–185, 2019.

[89] G. F. de Arruda, E. Cozzo, T. P. Peixoto, F. A. Rodrigues, and Y. Moreno. Disease localization in multilayer networks. *Physical Review X*, 7 (1): 011014, 2017.

[90] M. De Domenico. Diffusion geometry unravels the emergence of functional clusters in collective phenomena. *Physical Review Letters*, 118 (16): 168301, 2017.

[91] M. De Domenico. Multilayer modeling and analysis of human brain networks. *GigaScience*, 6 (5): 1–8, 2017.

[92] M. De Domenico. Multilayer network modeling of integrated biological systems. Comment on "Network science of biological systems at different scales: A review" by Gosak et al. *Physics of Life Reviews*, 2018.

[93] M. De Domenico. Multilayer Networks Illustrated, 2020. http://doi.org/10.17605/OSF.IO/GY53K. Accessed November 25, 2020.

[94] M. De Domenico. *Multilayer networks: Analysis and visualization. Introduction to muxViz with R*. Springer-Verlag, 2021.

[95] M. De Domenico and J. Biamonte. Spectral entropies as information-theoretic tools for complex network comparison. *Physical Review X*, 6 (4): 041062, 2016.

[96] M. De Domenico et al. Complexity explained. *OSF*, 2019. osf.io/tqgnw. Accessed November 25, 2020.

[97] M. De Domenico, A. Solé-Ribalta, E. Cozzo, et al. Mathematical formulation of multilayer networks. *Physical Review X*, 3 (4): 041022, 2013.

[98] M. De Domenico, A. Solé-Ribalta, S. Gómez, and A. Arenas. Navigability of interconnected networks under random failures. *PNAS*, 111 (23): 8351–8356, 2014.

[99] M. De Domenico, A. Lancichinetti, A. Arenas, and M. Rosvall. Identifying modular flows on multilayer networks reveals highly overlapping organization in interconnected systems. *Physical Review X*, 5 (1): 011027, 2015.

[100] M. De Domenico, V. Nicosia, A. Arenas, and V. Latora. Structural reducibility of multilayer networks. *Nature Communications*, 6: 6864, 2015.

[101] M. De Domenico, M. A. Porter, and A. Arenas. MuxViz: A tool for multilayer analysis and visualization of networks. *Journal of Complex Networks*, 3 (2): 159–176, 2015.

[102] M. De Domenico, A. Solé-Ribalta, E. Omodei, S. Gómez, and A. Arenas. Ranking in interconnected multilayer networks reveals versatile nodes. *Nature Communications*, 6: 6868, 2015.

[103] M. De Domenico, C. Granell, M. A. Porter, and A. Arenas. The physics of spreading processes in multilayer networks. *Nature Physics*, 12 (10): 901, 2016.

[104] M. De Domenico, S. Sasai, and A. Arenas. Mapping multiplex hubs in human functional brain networks. *Frontiers in Neuroscience*, 10: 326, 2016.

[105] C. I. Del Genio, J. Gómez-Gardeñes, I. Bonamassa, and S. Boccaletti. Synchronization in networks with multiple interaction layers. *Science Advances*, 2 (11): 1–10, 2016.

[106] R. M. del Rio-Chanona, Y. Korniyenko, M. Patnam, and M. A. Porter. The multiplex nature of global financial contagions. *Applied Network Science*, 5 (1): 1–23, 2020.

[107] F. Della Rossa, L. Pecora, K. Blaha, et al. Symmetries and cluster synchronization in multilayer networks. *Nature Communications*, 11 (1): 1–17, 2020.

[108] M. Diakonova, V. Nicosia, V. Latora, and M. San Miguel. Irreducibility of multilayer network dynamics: The case of the voter model. *New Journal of Physics*, 18 (2): 023010, 2016.

[109] M. Dickison, S. Havlin, and H. E. Stanley. Epidemics on interconnected networks. *Physical Review E*, 85 (6): 066109, 2012.

[110] S. N. Dorogovtsev, A. V. Goltsev, and J. F. F. Mendes. Critical phenomena in complex networks. *Reviews of Modern Physics*, 80 (4): 1275, 2008.

[111] M. Duh, M. Gosak, M. Slavinec, and M. Perc. Assortativity provides a narrow margin for enhanced cooperation on multilayer networks. *New Journal of Physics*, 21: 123016, 2019.

[112] D. Edler, L. Bohlin, and M. Rosvall. Mapping higher-order network flows in memory and multilayer networks with infomap. *Algorithms*, 10 (4): 112, 2017.

[113] A. V. Esquivel and M. Rosvall. Compression of flow can reveal overlapping-module organization in networks. *Physical Review X*, 1 (2): 021025, 2011.

[114] E. Estrada. *The structure of complex networks: Theory and applications*. Oxford University Press, 2012.

[115] E. Estrada. Communicability geometry of multiplexes. *New Journal of Physics*, 21 (1): 015004, 2019.

[116] E. Estrada and J. Gómez-Gardeñes. Communicability reveals a transition to coordinated behavior in multiplex networks. *Physical Review E*, 89 (4): 042819, 2014.

[117] EUROCONTROL. Ash-cloud of April and May 2010: Impact on air traffic, 2010. https://www.eurocontrol.int/publication/ash -cloud -april-and-may-2010-impact-air-traffic. Accessed March 17, 2020.

[118] S. Fortunato. Community detection in graphs. *Physics Reports*, 486 (3–5): 75–174, 2010.

[119] S. Fortunato and M. Barthelemy. Resolution limit in community detection. *PNAS*, 104 (1): 36–41, 2007.

[120] S. Fortunato and D. Hric. Community detection in networks: A user guide. *Physics Reports*, 659: 1–44, 2016.

[121] L. C. Freeman, S. P. Borgatti, and D. R. White. Centrality in valued graphs: A measure of betweenness based on network flow. *Social Networks*, 13 (2): 141–154, 1991.

[122] S. Funk, E. Gilad, C. Watkins, and V. A. A. Jansen. The spread of awareness and its impact on epidemic outbreaks. *PNAS*, 106 (16): 6872–6877, 2009.

[123] S. Funk, S. Bansal, C. T. Bauch, et al. Nine challenges in incorporating the dynamics of behaviour in infectious diseases models. *Epidemics*, 10: 21–25, 2015.

[124] E. Galimberti, F. Bonchi, F. Gullo, and T. Lanciano. Core decomposition in multilayer networks: Theory, algorithms, and applications. *ACM Transactions on Knowledge Discovery from Data (TKDD)*, 14 (1): 1–40, 2020.

[125] R. Gallotti and M. Barthelemy. Anatomy and efficiency of urban multimodal mobility. *Scientific Reports*, 4 (1): 1–9, 2014.

[126] R. Gallotti and M. Barthelemy. The multilayer temporal network of public transport in Great Britain. *Scientific Data*, 2 (1): 1–8, 2015.

[127] R. Gallotti, G. Bertagnolli, and M. De Domenico. Unraveling the hidden organisation of urban systems and their mobility flows. *EPJ Data Science*, 10 (1): 1–17, 2021.

[128] L. V. Gambuzza, M. Frasca, and J. Gómez-Gardeñes. Intra-layer synchronization in multiplex networks. *Europhysics Letters*, 110 (2): 20010, 2015.

[129] J. Gao, S. V. Buldyrev, H. E. Stanley, and S. Havlin. Networks formed from interdependent networks. *Nature Physics*, 8 (1): 40, 2012.

[130] L. Gauvin, A. Panisson, and C. Cattuto. Detecting the community structure and activity patterns of temporal networks: A non-negative tensor factorization approach. *PloS One*, 9 (1): e86028, 2014.

[131] A. Ghavasieh and M. De Domenico. Enhancing transport properties in interconnected systems without altering their structure. *Physical Review Research*, 2: 013155, 2020.

[132] A. Ghavasieh, C. Nicolini, and M. De Domenico. Statistical physics of complex information dynamics. *Physical Review E*, 102: 052304, 2020.

[133] M. Girvan and M. E. J. Newman. Community structure in social and biological networks. *PNAS*, 99 (12): 7821–7826, 2002.

[134] A. Goldenberg, A. X. Zheng, S. E. Fienberg, et al. A survey of statistical network models. *Foundations and Trends® in Machine Learning*, 2 (2): 129–233, 2010.

[135] N. Goldenfeld. *Lectures on phase transitions and the renormalization group*. CRC Press, 2018.

[136] S. Gomez, A. Diaz-Guilera, J. Gomez-Gardenes, et al. Diffusion dynamics on multiplex networks. *Physical Review Letters*, 110 (2): 028701, 2013.

[137] J. Gómez-Gardeñes, S. Gómez, A. Arenas, and Y. Moreno. Explosive synchronization transitions in scale-free networks. *Physical Review Letters*, 106 (12): 1–6, 2011.

[138] J. Gómez-Gardeñes, M. Romance, R. Criado, D. Vilone, and A. Sánchez. Evolutionary games defined at the network mesoscale: The public goods game. *Chaos*, 21 (1): 1–10, 2011.

[139] J. Gómez-Gardenes, I. Reinares, A. Arenas, and L. M. Floría. Evolution of cooperation in multiplex networks. *Scientific Reports*, 2: 620, 2012.

[140] J. Gomez-Gardenes, M. de Domenico, G. Gutierrez, A. Arenas, and S. Gomez. Layer-layer competition in multiplex complex networks. *Philosophical Transactions of the Royal Society A*, 373 (2056): 20150117, 2015.

[141] C. Granell, S. Gómez, and A. Arenas. Dynamical interplay between awareness and epidemic spreading in multiplex networks. *Physical Review Letters*, 111 (12), 2013.

[142] C. Granell, S. Gómez, and A. Arenas. Competing spreading processes on multiplex networks: Awareness and epidemics. *Physical Review E*, 90 (1): 012808, 2014.

[143] P. Grassberger. Percolation transitions in the survival of interdependent agents on multiplex networks, catastrophic cascades, and solid-on-solid surface growth. *Physical Review E*, 91 (6): 062806, 2015.

[144] R. Guimera and L. A. N. Amaral. Functional cartography of complex metabolic networks. *Nature*, 433 (7028): 895, 2005.

[145] A. Hackett, D. Cellai, S. Gómez, A. Arenas, and J. P. Gleeson. Bond percolation on multiplex networks. *Physical Review X*, 6 (2): 021002, 2016.

[146] A. Halu, R. J. Mondragón, P. Panzarasa, and G. Bianconi. Multiplex PageRank. *PloS One*, 8 (10): e78293, 2013.

[147] P. W. Holland, K. B. Laskey, and S. Leinhardt. Stochastic blockmodels: First steps. *Social Networks*, 5 (2): 109–137, 1983.

[148] P. Holme. Modern temporal network theory: A colloquium. *European Physical Journal B*, 88 (9): 234, 2015.

[149] P. Holme and J. Saramäki. Temporal networks. *Physics Reports*, 519 (3): 97–125, 2012.

[150] Y. Hu, D. Zhou, R. Zhang, et al. Percolation of interdependent networks with intersimilarity. *Physical Review E*, 88 (5): 052805, 2013.

[151] Y. Hu, S. Havlin, and H. A. Makse. Conditions for viral influence spreading through multiplex correlated social networks. *Physical Review X*, 4 (2): 021031, 2014.

[152] X. Huang, S. Shao, H. Wang, et al. The robustness of interdependent clustered networks. *Europhysics Letters*, 101 (1): 18002, 2013.

[153] X. Huang, I. Vodenska, S. Havlin, and H. E. Stanley. Cascading failures in bi-partite graphs: Model for systemic risk propagation. *Scientific Reports*, 3: 1219, 2013.

[154] J. Iacovacci, C. Rahmede, A. Arenas, and G. Bianconi. Functional multiplex PageRank. *Europhysics Letters*, 116 (2): 28004, 2016.

[155] S. Jalan and A. Singh. Cluster synchronization in multiplex networks. *Europhysics Letters*, 113 (3): 2–7, 2016.

[156] S. Jang, J. S. Lee, S. Hwang, and B. Kahng. Ashkin-Teller model and diverse opinion phase transitions on multiplex networks. *Physical Review E*, 92 (2): 022110, 2015.

[157] H. J. Jensen. *Self-organized criticality: Emergent complex behavior in physical and biological systems*, volume 10. Cambridge University Press, 1998.

[158] L. Katz. A new status index derived from sociometric analysis. *Psychometrika*, 18 (1): 39–43, 1953.

[159] D. Kempe, J. Kleinberg, and É. Tardos. Maximizing the spread of influence through a social network. In *Proceedings of the 9th ACM SIGKDD International Conference on Knowledge Discovery and Data Mining*, pages 137–146, 2003.

[160] D. Y. Kenett, J. Gao, X. Huang, et al. Network of interdependent networks: Overview of theory and applications. In *Networks of Networks: The Last Frontier of Complexity*, pages 3–36. Springer, 2014.

[161] M. Kivelä, A. Arenas, M. Barthelemy, et al. Multilayer networks. *Journal of Complex Networks*, 2 (3): 203–271, 2014.

[162] J. M. Kleinberg. Authoritative sources in a hyperlinked environment. *Journal of the ACM*, 46 (5): 604–632, 1999.

[163] K. K. Kleineberg and D. Helbing. Topological enslavement in evolutionary games on correlated multiplex networks. *New Journal of Physics*, 20 (5), 2018.

[164] T. G. Kolda and B. W. Bader. Tensor decompositions and applications. *SIAM Review*, 51 (3): 455–500, 2009.

[165] N. E. Kouvaris, S. Hata, and A. Díaz-Guilera. Pattern formation in multiplex networks. *Scientific Reports*, 5 (1): 1–9, 2015.

[166] I. Kryven. Bond percolation in coloured and multiplex networks. *Nature Communications*, 10 (1): 1–16, 2019.

[167] L. Lacasa, I. P. Mariño, J. Miguez, et al. Multiplex decomposition of non-Markovian dynamics and the hidden layer reconstruction problem. *Physical Review X*, 8 (3): 031038, 2018.

[168] L. Lacasa, S. Stramaglia, and D. Marinazzo. Beyond pairwise network similarity: Exploring mediation and suppression between networks. *Communications Physics*, 4 (1): 1–8, 2021.

[169] R. Lambiotte, J.-C. Delvenne, and M. Barahona. Random walks, Markov processes and the multiscale modular organization of complex networks. *IEEE Transactions on Network Science and Engineering*, 1 (2): 76–90, 2014.

[170] R. Lambiotte, M. Rosvall, and I. Scholtes. From networks to optimal higher-order models of complex systems. *Nature Physics*, 15 (4): 313–320, 2019.

[171] V. Latora and M. Marchiori. Efficient behavior of small-world networks. *Physical Review Letters*, 87 (19): 198701, 2001.

[172] V. Latora and M. Marchiori. Economic small-world behavior in weighted networks. *European Physical Journal B*, 32 (2): 249–263, 2003.

[173] V. Latora, V. Nicosia, and G. Russo. *Complex networks: Principles, methods and applications*. Cambridge University Press, 2017.

[174] K.-M. Lee, K.-I. Goh, and I.-M. Kim. Sandpiles on multiplex networks. *Journal of the Korean Physical Society*, 60 (4): 641–647, 2012.

[175] K.-M. Lee, J. Y. Kim, W.-k. Cho, K.-I. Goh, and I. M. Kim. Correlated multiplexity and connectivity of multiplex random networks. *New Journal of Physics*, 14 (3): 033027, 2012.

[176] K.-M. Lee, B. Min, and K.-I. Goh. Towards real-world complexity: An introduction to multiplex networks. *European Physical Journal B*, 88 (2): 48, 2015.

[177] E. A. Leicht and R. M. D'Souza. Percolation on interacting networks. *arXiv:0907.0894*, 2009.

[178] I. Leyva, A. Navas, I. Sendiña-Nadal, et al. Explosive transitions to synchronization in networks of phase oscillators. *Scientific Reports*, 3: 1–5, 2013.

[179] I. Leyva, I. Sendiña-Nadal, R. Sevilla-Escoboza, et al. Relay synchronization in multiplex networks. *Scientific Reports*, 8, 2018.

[180] W. Li, A. Bashan, S. V. Buldyrev, H. E. Stanley, and S. Havlin. Cascading failures in interdependent lattice networks: The critical role of the length of dependency links. *Physical Review Letters*, 108 (22): 228702, 2012.

[181] A. Lima, M. De Domenico, V. Pejovic, and M. Musolesi. Exploiting cellular data for disease containment and information campaigns strategies in country-wide epidemics. In *Proc. of 3rd Intern. Conf. on the Analysis of Mobile Phone Datasets, Boston, USA*, page 1. NETMOB, 2013.

[182] A. Lima, M. De Domenico, V. Pejovic, and M. Musolesi. Disease containment strategies based on mobility and information dissemination. *Scientific Reports*, 5: 10650, 2015.

[183] R. Louf and M. Barthelemy. Patterns of residential segregation. *PloS One*, 11 (6): e0157476, 2016.

[184] M. Magnani, B. Micenkova, and L. Rossi. Combinatorial analysis of multiple networks. *arXiv:1303.4986*, 2013.

[185] N. Mantel. The detection of disease clustering and a generalized regression approach. *Cancer Research*, 27 (2 Part 1): 209–220, 1967.

[186] E. A. Martens, E. Barreto, S. H. Strogatz, et al. Exact results for the Kuramoto model with a bimodal frequency distribution. *Physical Review E*, 79 (2): 026204, 2009.

[187] E. Massaro and F. Bagnoli. Epidemic spreading and risk perception in multiplex networks: A self-organized percolation method. *Physical Review E*, 90 (5): 052817, 2014.

[188] N. Masuda, M. A. Porter, and R. Lambiotte. Random walks and diffusion on networks. *Physics Reports*, 2017.

[189] J. T. Matamalas, J. Poncela-Casasnovas, S. Gómez, and A. Arenas. Strategical incoherence regulates cooperation in social dilemmas on multiplex networks. *Scientific Reports*, 5: 9519, 2015.

[190] G. Menichetti, D. Remondini, P. Panzarasa, R. J. Mondragón, and G. Bianconi. Weighted multiplex networks. *PloS One*, 9 (6): e97857, 2014.

[191] A. B. Migliano, A. E. Page, J. Gómez-Gardeñes, et al. Characterization of hunter-gatherer networks and implications for cumulative culture. *Nature Human Behaviour*, 1 (2): 1–6, 2017.

[192] B. Min and K.-I. Goh. Multiple resource demands and viability in multiplex networks. *Physical Review E*, 89 (4): 040802, 2014.

[193] B. Min, S. Do Yi, K.-M. Lee, and K.-I. Goh. Network robustness of multiplex networks with interlayer degree correlations. *Physical Review E*, 89 (4): 042811, 2014.

[194] B. Min, S. Lee, K.-M. Lee, and K.-I. Goh. Link overlap, viability, and mutual percolation in multiplex networks. *Chaos, Solitons & Fractals*, 72: 49–58, 2015.

[195] M. Molloy and B. Reed. A critical point for random graphs with a given degree sequence. *Random Structures & Algorithms*, 6 (2–3): 161–180, 1995.

[196] R. G. Morris and M. Barthelemy. Transport on coupled spatial networks. *Physical Review Letters*, 109 (12): 128703, 2012.

[197] A. E. Motter and Y.-C. Lai. Cascade-based attacks on complex networks. *Physical Review E*, 66 (6): 065102, 2002.

[198] P. J. Mucha, T. Richardson, K. Macon, M. A. Porter, and J.-P. Onnela. Community structure in time-dependent, multiscale, and multiplex networks. *Science*, 328 (5980): 876–878, 2010.

[199] D. R. Nelson. Recent developments in phase transitions and critical phenomena. *Nature*, 269 (5627): 379–383, 1977.

[200] North American Electric Reliability Council Steering Group. Technical a Analysis of the August 14, 2003, blackout: What happened, why, and what did we learn? Technical report, NERC, 2004. Report to the North American Electric Reliability Council Board of Trustees.

[201] M. E. J. Newman. Modularity and community structure in networks. *PNAS*, 103 (23): 8577–8582, 2006.

[202] M. E. J. Newman. Communities, modules and large-scale structure in networks. *Nature Physics*, 8 (1): 25, 2012.

[203] M. E. J. Newman. *Networks*. Oxford University Press, 2018.

[204] M. E. J. Newman, S. H. Strogatz, and D. J. Watts. Random graphs with arbitrary degree distributions and their applications. *Physical Review E*, 64 (2): 026118, 2001.

[205] V. Nicosia and V. Latora. Measuring and modeling correlations in multiplex networks. *Physical Review E*, 92 (3): 032805, 2015.

[206] V. Nicosia, G. Bianconi, V. Latora, and M. Barthelemy. Growing multiplex networks. *Physical Review Letters*, 111: 058701, 2013.

[207] V. Nicosia, M. Valencia, M. Chavez, A. Díaz-Guilera, and V. Latora. Remote synchronization reveals network symmetries and functional modules. *Physical Review Letters*, 110 (17): 1–5, 2013.

[208] V. Nicosia, P. S. Skardal, A. Arenas, and V. Latora. Collective phenomena emerging from the interactions between dynamical processes in multiplex networks. *Physical Review Letters*, 118 (13): 138302, 2017.

[209] J. D. Noh and H. Rieger. Random walks on complex networks. *Physical Review Letters*, 92 (11): 118701, 2004.

[210] North American Electric Reliability Council. 1996 system disturbances. Review of selected 1996 electric system disturbances in North America. Technical report, North American Electric Reliability Council, 2002.

[211] M. A. Nowak and R. M. May. Evolutionary games and spatial chaos. *Nature*, 359: 826–829, 1992.

[212] M. A. Nowak, C. E. Tarnita, and T. Antal. Evolutionary dynamics in structured populations. *Philosophical Transactions of the Royal Society B*, 365 (1537): 19–30, 2010.

[213] K. Nowicki and T. A. B. Snijders. Estimation and prediction for stochastic blockstructures. *Journal of the American Statistical Association*, 96 (455): 1077–1087, 2001.

[214] S. Osat, A. Faqeeh, and F. Radicchi. Optimal percolation on multiplex networks. *Nature Communications*, 8 (1): 1540, 2017.

[215] L. Page, S. Brin, R. Motwani, and T. Winograd. The PageRank citation ranking: Bringing order to the Web. Technical report, Stanford InfoLab, 1999.

[216] A. R. Pamfil, S. D. Howison, R. Lambiotte, and M. A. Porter. Relating modularity maximization and stochastic block models in multilayer networks. *SIAM Journal on Mathematics of Data Science*, 1 (4): 667–698, 2019.

[217] A. R. Pamfil, S. D. Howison, and M. A. Porter. Inference of edge correlations in multilayer networks. *Physical Review E*, 102 (6): 062307, 2020.

[218] R. Parshani, S. V. Buldyrev, and S. Havlin. Interdependent networks: Reducing the coupling strength leads to a change from a first to second order percolation transition. *Physical Review Letters*, 105 (4): 048701, 2010.

[219] L. M. Pecora and T. L. Carroll. Master stability functions for synchronized coupled systems. *Physical Review Letters*, 80 (10): 2109–2112, 1998.

[220] L. M. Pecora, F. Sorrentino, A. M. Hagerstrom, T. E. Murphy, and R. Roy. Cluster synchronization and isolated desynchronization in complex networks with symmetries. *Nature Communications*, 5 (May), 2014.

[221] T. P. Peixoto. Inferring the mesoscale structure of layered, edge-valued, and time-varying networks. *Physical Review E*, 92 (4): 042807, 2015.

[222] T. P. Peixoto. Nonparametric Bayesian inference of the microcanonical stochastic block model. *Physical Review E*, 95 (1): 012317, 2017.

[223] T. P. Peixoto. Bayesian stochastic blockmodeling. In Advances in network clustering and blockmodeling, Wiley, pages 289–332. 2019.

[224] M. Perc, J. J. Jordan, D. G. Rand, et al. Statistical physics of human cooperation. *Physics Reports*, 687: 1–51, 2017.

[225] S. Pilosof, M. A. Porter, M. Pascual, and S. Kéfi. The multilayer nature of ecological networks. *Nature Ecology & Evolution*, 1 (4): 0101, 2017.

[226] P. Pons and M. Latapy. Computing communities in large networks using random walks. *Journal of Graph Algorithms and Applications*, 10 (2): 191–218, 2006.

[227] M. Pósfai, J. Gao, S. P. Cornelius, A.-L. Barabási, and R. M. D'Souza. Controllability of multiplex, multi-time-scale networks. *Physical Review E*, 94 (3): 032316, 2016.

[228] M. Pósfai, N. Braun, B. A. Beisner, B. McCowan, and R. M. D'Souza. Consensus ranking for multi-objective interventions in multiplex networks. *New Journal of Physics*, 21 (5): 055001, 2019.

[229] T. Qin and K. Rohe. Regularized spectral clustering under the degree-corrected stochastic blockmodel. In *Advances in neural information processing systems*, vol. 2, pages 3120–3128, Curran Associates, 2013.

[230] F. Radicchi. Percolation in real interdependent networks. *Nature Physics*, 11 (7): 597–602, 2015.

[231] F. Radicchi and A. Arenas. Abrupt transition in the structural formation of interconnected networks. *Nature Physics*, 9 (11): 717, 2013.

[232] F. Radicchi and G. Bianconi. Redundant interdependencies boost the robustness of multiplex networks. *Physical Review X*, 7: 011013, 2017.

[233] R. Ramezanian, M. Magnani, M. Salehi, and D. Montesi. Diffusion of innovations over multiplex social networks. In *International Symposium on Artificial Intelligence and Signal Processing (AISP)*, pages 300–304. Institute of Electrical and Electronics Engineers, 2015.

[234] S. D. S. Reis, Y. Hu, A. Babino, et al. Avoiding catastrophic failure in correlated networks of networks. *Nature Physics*, 10 (10): 762, 2014.

[235] R. J. Requejo and A. Díaz-Guilera. Replicator dynamics with diffusion on multiplex networks. *Physical Review E*, 94 (2): 022301, 2016.

[236] V. Rosato, L. Issacharoff, F. Tiriticco, et al. Modelling interdependent infrastructures using interacting dynamical models. *International Journal of Critical Infrastructures*, 4 (1–2): 63–79, 2008.

[237] M. Rosvall and C. T. Bergstrom. An information-theoretic framework for resolving community structure in complex networks. *PNAS*, 104 (18): 7327–7331, 2007.

[238] M. Rosvall and C. T. Bergstrom. Maps of random walks on complex networks reveal community structure. *PNAS*, 105 (4): 1118–1123, 2008.

[239] M. Rosvall, A. V. Esquivel, A. Lancichinetti, J. D. West, and R. Lambiotte. Memory in network flows and its effects on spreading dynamics and community detection. *Nature Communications*, 5: 4630, 2014.

[240] M. Rubinov and O. Sporns. Complex network measures of brain connectivity: Uses and interpretations. *Neuroimage*, 52 (3): 1059–1069, 2010.

[241] A. Saa. Symmetries and synchronization in multilayer random networks. *Physical Review E*, 97 (4): 042304, 2018.

[242] F. D. Sahneh, C. Scoglio, and P. Van Mieghem. Generalized epidemic mean-field model for spreading processes over multilayer complex networks. *IEEE/ACM Transactions on Networking (TON)*, 21 (5): 1609–1620, 2013.

[243] M. Salehi, R. Sharma, M. Marzolla, et al. Spreading processes in multilayer networks. *IEEE Transactions on Network Science and Engineering*, 2 (2): 65–83, 2015.

[244] V. Salnikov, M. T. Schaub, and R. Lambiotte. Using higher-order Markov models to reveal flow-based communities in networks. *Scientific Reports*, 6: 23194, 2016.

[245] A. Santoro and V. Nicosia. Optimal percolation in correlated multilayer networks with overlap. *Physical Review Research*, 2 (3): 033122, 2020.

[246] A. Santoro, V. Latora, G. Nicosia, and V. Nicosia. Pareto optimality in multilayer network growth. *Physical Review Letters*, 121 (12): 128302, 2018.

[247] F. C. Santos, J. M. Pacheco, and T. Lenaerts. Evolutionary dynamics of social dilemmas in structured heterogeneous populations. *PNAS*, 103 (9): 3490–3494, 2006.

[248] F. C. Santos, M. D. Santos, and J. M. Pacheco. Social diversity promotes the emergence of cooperation in public goods games. *Nature*, 454 (7201): 213–216, 2008.

[249] J. Sanz, C.-Y. Xia, S. Meloni, and Y. Moreno. Dynamics of interacting diseases. *Physical Review X*, 4 (4): 041005, 2014.

[250] C. M. Schneider, N. Yazdani, N. A. M. Araújo, S. Havlin, and H. J. Herrmann. Towards designing robust coupled networks. *Scientific Reports*, 3: 1969, 2013.

[251] J. Scott. Popularity, mediation and exclusion. *In Social network analysis*, pages 95–112, Sage, 2017.

[252] S. B. Seidman. Network structure and minimum degree. *Social Networks*, 5 (3): 269–287, 1983.

[253] L. M. Shekhtman, M. M. Danziger, and S. Havlin. Recent advances on failure and recovery in networks of networks. *Chaos, Solitons & Fractals*, 90: 28–36, 2016.

[254] A. Singh, S. Ghosh, S. Jalan, and J. Kurths. Synchronization in delayed multiplex networks. *Europhysics Letters*, 111 (3): 30010, 2015.

[255] P. S. Skardal and A. Arenas. Control of coupled oscillator networks with application to microgrid technologies. *Science Advances*, 1 (7): e1500339, 2015.

[256] T. A. B. Snijders and K. Nowicki. Estimation and prediction for stochastic blockmodels for graphs with latent block structure. *Journal of Classification*, 14 (1): 75–100, 1997.

[257] L. Solá, M. Romance, R. Criado, et al. Eigenvector centrality of nodes in multiplex networks. *Chaos*, 23 (3): 033131, 2013.

[258] A. Sole-Ribalta, M. De Domenico, N. E. Kouvaris, et al. Spectral properties of the Laplacian of multiplex networks. *Physical Review E*, 88 (3): 032807, 2013.

[259] A. Solé-Ribalta, M. De Domenico, S. Gómez, and A. Arenas. Centrality rankings in multiplex networks. In *Proceedings of the 2014 ACM Conference on Web Science*, pages 149–155. Association for Computing Machinery, 2014.

[260] A. Solé-Ribalta, M. De Domenico, S. Gómez, and A. Arenas. Random walk centrality in interconnected multilayer networks. *Physica D*, 323: 73–79, 2016.

[261] A. Solé-Ribalta, S. Gómez, and A. Arenas. Congestion induced by the structure of multiplex networks. *Physical Review Letters*, 116 (10): 108701, 2016.

[262] S.-W. Son, P. Grassberger, and M. Paczuski. Percolation transitions are not always sharpened by making networks interdependent. *Physical Review Letters*, 107 (19): 195702, 2011.

[263] S.-W. Son, G. Bizhani, C. Christensen, P. Grassberger, and M. Paczuski. Percolation theory on interdependent networks based on epidemic spreading. *Europhysics Letters*, 97 (1): 16006, 2012.

[264] D. Soriano-Paños, L. Lotero, A. Arenas, and J. Gómez-Gardeñes. Spreading processes in multiplex metapopulations containing different Mobility networks. *Physical Review X*, 8 (3): 031039, 2018.

[265] F. Sorrentino, L. M. Pecora, A. M. Hagerstrom, T. E. Murphy, and R. Roy. Complete characterization of the stability of cluster synchronization in complex dynamical networks. *Science Advances*, 2 (4): 1–9, 2016.

[266] O. Sporns. Network attributes for segregation and integration in the human brain. *Current Opinion in Neurobiology*, 23 (2): 162–171, 2013.

[267] H. E. Stanley. Scaling, universality, and renormalization: Three pillars of modern critical phenomena. *Reviews of Modern Physics*, 71 (2): S358, 1999.

[268] D. Stauffer and A. Aharony. *Introduction to percolation theory*. CRC Press, 2018.

[269] M. Stella, N. M. Beckage, M. Brede, and M. De Domenico. Multiplex model of mental lexicon reveals explosive learning in humans. *Scientific Reports*, 8 (1): 2259, 2018.

[270] S. H. Strogatz. From Kuramoto to Crawford: Exploring the onset of synchronization in populations of coupled oscillators. *Physica D*, 143 (1–4): 1–20, 2000.

[271] F. Tan, J. Wu, Y. Xia, and K. T. Chi. Traffic congestion in interconnected complex networks. *Physical Review E*, 89 (6): 062813, 2014.

[272] D. Taylor, S. Shai, N. Stanley, and P. J. Mucha. Enhanced detectability of community structure in multilayer networks through layer aggregation. *Physical Review Letters*, 116 (22): 228301, 2016.

[273] D. Taylor, R. S. Caceres, and P. J. Mucha. Super-resolution community detection for layer-aggregated multilayer networks. *Physical Review X*, 7 (3): 031056, 2017.

[274] D. Taylor, M. A. Porter, and P. J. Mucha. Tunable eigenvector-based centralities for multiplex and temporal networks. *Multiscale Modeling & Simulation*, 19 (1): 113–147, 2021.

[275] A. Tejedor, A. Longjas, E. Foufoula-Georgiou, T. T. Georgiou, and Y. Moreno. Diffusion dynamics and optimal coupling in multiplex networks with directed layers. *Physical Review X*, 8 (3): 031071, 2018.

[276] P. Tewarie, A. Hillebrand, B. W. van Dijk, et al. Integrating cross-frequency and within band functional networks in resting-state meg: A multi-layer network approach. *Neuroimage*, 142: 324–336, 2016.

[277] S. Timóteo, M. Correia, S. Rodríguez-Echeverría, H. Freitas, and R. Heleno. Multilayer networks reveal the spatial structure of seed-dispersal interactions across the great rift landscapes. *Nature Communications*, 9 (1): 140, 2018.

[278] S. Torreggiani, G. Mangioni, M. J. Puma, and G. Fagiolo. Identifying the community structure of the food-trade international multi-network. *Environmental Research Letters*, 13 (5): 054026, 2018.

[279] V. A. Traag. Complex contagion of campaign donations. *PloS One*, 11 (4): e0153539, 2016.

[280] A. Trewavas. A brief history of systems biology: *"Every object that biology studies is a system of systems."* Francois Jacob (1974). *The Plant Cell*, 18 (10): 2420–2430, 2006.

[281] L. R. Tucker. Some mathematical notes on three-mode factor analysis. *Psychometrika*, 31 (3): 279–311, 1966.

[282] E. Valdano, L. Ferreri, C. Poletto, and V. Colizza. Analytical computation of the epidemic threshold on temporal networks. *Physical Review X*, 5 (2): 021005, 2015.

[283] A. Valdeolivas, L. Tichit, C. Navarro, et al. Random walk with restart on multiplex and heterogeneous biological networks. *Bioinformatics*, 2018.

[284] L. D. Valdez, L. Shekhtman, C. E. La Rocca, et al. Cascading failures in complex networks. *Journal of Complex Networks*, 8 (2): cnaa013, 2020.

[285] T. Valles-Catala, F. A. Massucci, R. Guimera, and M. Sales-Pardo. Multilayer stochastic block models reveal the multilayer structure of complex networks. *Physical Review X*, 6 (1): 011036, 2016.

[286] F. Velásquez-Rojas. Interacting opinion and disease dynamics in multiplex networks: Discontinuous phase transition and nonmonotonic consensus times. *Physical Review E*, 95 (5): 052315, 2017.

[287] R. Vermeulen, E. L. Schymanski, A.-L. Barabási, and G. W. Miller. The exposome and health: Where chemistry meets biology. *Science*, 367 (6476): 392–396, 2020.

[288] N. Verstraete, G. Jurman, G. Bertagnolli, et al. CovMulNet19, integrating proteins, diseases, drugs, and symptoms: A network medicine approach to COVID-19. *Network and Systems Medicine*, 3 (1): 130–141, 2020.

[289] A. Vespignani. Complex networks: The fragility of interdependency. *Nature*, 464 (7291): 984, 2010.

[290] I. Voitalov, P. van der Hoorn, M. Kitsak, F. Papadopoulos, and D. Krioukov. Weighted hypersoft configuration model. *Physical Review Research*, 2: 043157, 2020.

[291] H. Wang, Q. Li, G. D'Agostino, et al. Effect of the interconnected network structure on the epidemic threshold. *Physical Review E*, 88 (2): 022801, 2013.

[292] X. Wang, W. Li, L. Liu, et al. Promoting information diffusion through interlayer recovery processes in multiplex networks. *Physical Review E*, 96 (3): 032304, 2017.

[293] Z. Wang, L. Wang, and M. Perc. Degree mixing in multilayer networks impedes the evolution of cooperation. *Physical Review E*, 89: 052813, 2014.

[294] Z. Wang, M. A. Andrews, Z.-X. Wu, L. Wang, and C. T. Bauch. Coupled disease–behavior dynamics on complex networks: A review. *Physics of Life Reviews*, 15: 1–29, 2015.

[295] Z. Wang, L. Wang, A. Szolnoki, and M. Perc. Evolutionary games on multilayer networks: A colloquium. *European Physical Journal B*, 88 (5): 1–15, 2015.

[296] D. J. Watts. A simple model of global cascades on random networks. *PNAS*, 99 (9): 5766–5771, 2002.

[297] D. J. Watts and S. H. Strogatz. Collective dynamics of small-world networks. *Nature*, 393 (6684): 440, 1998.

[298] B. J. Williamson, M. De Domenico, and D. S. Kadis. Multilayer connector hub mapping reveals key brain regions supporting expressive language. *Brain Connectivity*, 11 (1): 45–55, 2021.

[299] H. Wu, R. G. James, and R. M. D'Souza. Correlated structural evolution within multiplex networks. *Journal of Complex Networks*, 8 (2): cnaa014, 2020.

[300] Q. Wu, X. Fu, M. Small, and X.-J. Xu. The impact of awareness on epidemic spreading in networks. *Chaos*, 22 (1): 013101, 2012.

[301] O. Yagan and V. Gligor. Analysis of complex contagions in random multiplex networks. *Physical Review E*, 86 (3): 036103, 2012.

[302] H. Yamamoto, S. Moriya, K. Ide, et al. Impact of modular organization on dynamical richness in cortical networks. *Science Advances*, 4 (11): eaau4914, 2018.

[303] Z. Yuan, C. Zhao, W.-X. Wang, Z. Di, and Y.-C. Lai. Exact controllability of multiplex networks. *New Journal of Physics*, 16 (10): 103036, 2014.

[304] W. W. Zachary. An information flow model for conflict and fission in small groups. *Journal of Anthropological Research*, 33 (4): 452–473, 1977.

[305] X. Zhang, S. Boccaletti, S. Guan, and Z. Liu. Explosive synchronization in adaptive and multilayer networks. *Physical Review Letters*, 114 (3): 1–5, 2015.

[306] Y. Zhang, V. Latora, and A. E. Motter. Unified treatment of dynamical processes on generalized networks: Higher-order, multilayer, and temporal interactions. *arXiv:2010.00613*, 2020.

[307] D.-W. Zhao, L.-H. Wang, Y.-F. Zhi, J. Zhang, and Z. Wang. The robustness of multiplex networks under layer node-based attack. *Scientific Reports*, 6: 24304, 2016.

[308] K. Zhao and G. Bianconi. Percolation on interacting, antagonistic networks. *Journal of Statistical Mechanics*, 2013 (05): P05005, 2013.

[309] O. Artime and M. De Domenico. From the origin of life to pandemics: Emergent phenomena in complex systems. *Philosophical Transactions of the Royal Society A*, 380 (2227): 20200410, 2022.

Cambridge Elements ≡

The Structure and Dynamics of Complex Networks

Guido Caldarelli

Ca' Foscari University of Venice

Guido Caldarelli is Full Professor of Theoretical Physics at Ca' Foscari University of Venice. Guido Caldarelli received his PhD from SISSA, after which he held postdoctoral positions in the Department of Physics and School of Biology, University of Manchester, and the Theory of Condensed Matter Group, University of Cambridge. He also spent some time at the University of Fribourg in Switzerland, at École Normale Supérieure in Paris, and at the University of Barcelona. His main scientific activity (interest?) is the study of networks, mostly analysis and modelling, with applications from financial networks to social systems as in the case of disinformation. He is the author of more than 200 journal publications on the subject, and three books, and is the current President of the Complex Systems Society (2018 to 2021).

About the Series

This cutting-edge new series provides authoritative and detailed coverage of the underlying theory of complex networks, specifically their structure and dynamical properties. Each Element within the series will focus upon one of three primary topics: static networks, dynamical networks, and numerical/computing network resources.

Cambridge Elements ≡

The Structure and Dynamics of Complex Networks

Elements in the Series

A full series listing is available at: www.cambridge.org/SDCN

Printed in the United States
by Baker & Taylor Publisher Services

Printed in the United States
by Baker & Taylor Publisher Services